A SCIENTIFIC WAY OF WAR

Studies in War, Society, and the Military

GENERAL EDITORS

Peter Maslowski
University of Nebraska–Lincoln

David Graff
Kansas State University

Reina Pennington
Norwich University

EDITORIAL BOARD

D'Ann Campbell
Director of Government and Foundation Relations
U.S. Coast Guard Foundation

Mark A. Clodfelter
National War College

Brooks D. Simpson
Arizona State University

Roger J. Spiller
George C. Marshall Professor of Military History
U.S. Army Command and General Staff College (retired)

Timothy H. E. Travers
University of Calgary

Arthur Waldron
Lauder Professor of International Relations
University of Pennsylvania

"Hope demonstrates that the science of military thought and theory during this period was about much more than simply preparing for and waging continental war."
—Andrew J. Ziebell, *Army History*

"A well-researched and well-written contribution to the early development of American military thought. Readers who are interested in West Point and the essential role that its graduates played in both the Mexican and Civil Wars will find the book to be especially interesting."
—Roger Cunningham, *Journal of America's Military Past*

"*A Scientific Way of War* will appeal to both professionals and laypersons with a serious interest in the U.S. Army, its premier professional academy, nineteenth-century American defense policy, the nature of a particular national approach to military theory and doctrine, and the professionalization of the American armed forces."
—Richard Swain, *Michigan War Studies Review*

"A detailed, thoughtful, and provocative explanation of the evolution of the U.S. Army's understanding of military science and why this scientific view of war was so important in the nation's military history and to the conduct of the Civil War."
—Brian McAllister Linn, Ralph R. Thomas Professor in Liberal Arts at Texas A&M University and author of *The Echo of Battle: The Army's Way of War*

"Highly recommended to any reader interested in the early development of the U.S. army."
—*Civil War Books and Authors*

"Truly original. . . . No other scholar has so successfully explained what Americans understood by the phrase 'military science' as taught—and modified over time—at West Point, and how that doctrine related to the nation's geographic position, quest for internal development, and preparation for and perceptions of war."

—Peter Maslowski, professor emeritus of history at the University of Nebraska—Lincoln and coauthor of *Looking for a Hero: Joe Ronnie Hooper and the Vietnam War*

"[Ian Hope's] keen insights and original interpretations come through clearly in his new book, *A Scientific Way of War*. His penetrating analyses revolutionize our understanding of American military thinking in the antebellum era. This book is required reading for anyone who would understand generalship and high command in the American Civil War."

—Richard J. Sommers, senior historian emeritus, U.S. Army Heritage and Education Center, U.S. Army War College

A SCIENTIFIC WAY OF WAR

Antebellum Military Science, West Point, and the Origins of American Military Thought

IAN C. HOPE

University of Nebraska Press
Lincoln

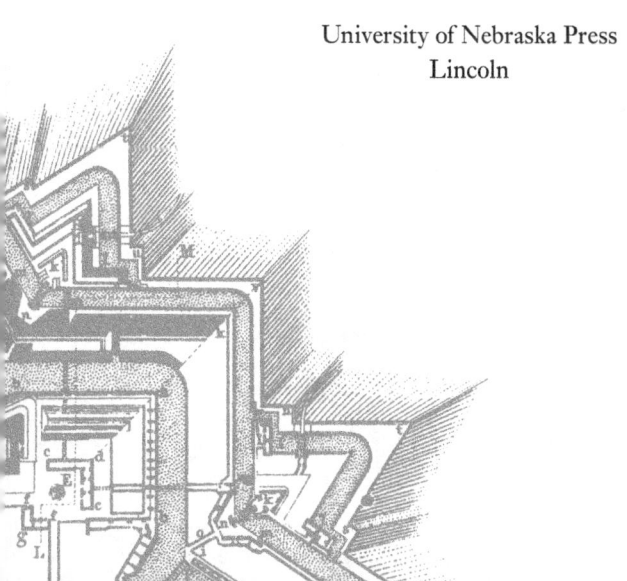

© 2015 by the Board of Regents of the University of Nebraska

All rights reserved
The University of Nebraska Press is part of a land-grant institution with campuses and programs on the past, present, and future homelands of the Pawnee, Ponca, Otoe-Missouria, Omaha, Dakota, Lakota, Kaw, Cheyenne, and Arapaho Peoples, as well as those of the relocated Ho-Chunk, Sac and Fox, and Iowa Peoples

Library of Congress Cataloging-in-Publication Data
Hope, Ian C.
A scientific way of war: antebellum military science, West Point, and the origins of American military thought / Ian C. Hope.
pages   cm.—(Studies in war, society, and the military)
Includes bibliographical references and index.
ISBN 978-0-8032-7685-7 (cloth: alk. paper)
ISBN 978-1-4962-3055-3 (paperback)
ISBN 978-0-8032-7716-8 (epub)
ISBN 978-0-8032-7717-5 (mobi)
ISBN 978-0-8032-7718-2 (pdf)
1. Military art and science—United States—History—19th century. 2. United States Military Academy—History—19th century.
3. Military education—United States—History—19th century. 4. United States Army—Officers—Training of—History—19th century. I. Title.
U43.U4H66 2015
355.00973'09034—dc23
2015009354

Set in Ehrhardt by L. Auten.

CONTENTS

List of Illustrations . . vii
Acknowledgments . . ix
Introduction . . 1
1. Colonial and Early National Military Science . . 17
2. Army Reforms, 1815–1820 . . 47
3. West Point's Scientific Curriculum . . 77
4. Internal Improvements . . 107
5. Jacksonian Military Science . . 129
6. Military Science during and after the Mexican War . . 161
7. Antebellum Military Science . . 183
8. Military Science in the Civil War . . 213
Conclusion . . 245

Appendix of Tables . . 255
Notes . . 277
Bibliography . . 311
Index . . 325

ILLUSTRATIONS

FIGURES

1. Von Bülow's optimum base of operations . . 42
2. War Department staff bureaus and key relationships . . 63
3. Vauban-style fortifications with trace showing concentration of flanking fire . . 94
4. O'Connor's reproduction of Jomini's "Configuration of a Theater of War" . . 100
5. Mahan's Theory of Strategy, the Science of Movement, ca. 1840 . . 155
6. The evolution of Mahan's ideas of strategy . . 202
7. Confederate fortifications near Centreville, Virginia, March 1862 . . 234
8. Confederate fortifications, with *chevaux-de-frise* beyond, at Petersburg, Virginia . . 235
9. City Point wharf with Federal artillery train, 1864–1865 . . 235
10. Yorktown, Virginia. Embarkation point for White House Landing . . 236
11. Pontoon bridge on the James River, 1864 . . 237
12. Wheeler's concept of strategic movements, ca. 1890 . . 252

MAPS

1. First System infrastructure, ca. 1800 . . 32
2. Second System infrastructure, ca. 1812 . . 39
3. Third System infrastructure, ca. 1821 . . 58
4. Major military roads, 1818 to 1840 . . 120
5. Strategic lines and points of the Mexican War . . 165
6. Third System sites and military posts, ca. 1855 . . 191
7. Military departments, December 1861 . . 220
8. Frontiers, depots, bases of operation, and Third System works, 1861–1863 . . 222
9. Western theater of operations . . 233
10. Eastern theater of operations . . 239
11. Grant's use of bases of operation, 1865 . . 241

TABLES

1. Regulations of the U.S. Military Academy at West Point, New York, 1823 . . 255
2. Regulations of the U.S. Military Academy at West Point, New York, 1832 . . 257
3. Regulations of the U.S. Military Academy at West Point, New York, 1856 . . 258
4. Officers performing staff functions, 1835 . . 259
5. Officers employed in staff functions during the antebellum period . . 260
6. Actual and possible membership of the Napoleon Club . . 261
7. Participants in Mahan's advanced engineer studies program . . 271
8. West Point graduates during the Civil War and their prewar experience . . 275

ACKNOWLEDGMENTS

This book started as a dissertation concerning Professor Dennis Hart Mahan and Jominian thought at antebellum West Point. The research failed to substantiate my original thesis. Evidence instead forced me to acknowledge that the real intellectual paradigm of the antebellum era was derivative of a doctrine called *military science*, a discovery that took me years to come to grips with as I produced first a dissertation and then this book. During this period of research and writing, I incurred significant debt to those who have inspired, guided, and sometimes driven me, and to those whose love, support, and patience sustained the process. If this project is in any way flawed, the fault is mine alone. If successful to any degree, however, it is because of the guidance and wisdom of Drs. Allan D. English, Roger Spiller, and Jane Errington, who for seven years (which included two combat tours in Afghanistan) waited for me to figure out what it was that I was trying to produce. Throughout the process I remained inspired by other great historians who have personally instructed and guided me, including Drs. James Stokesbury, Roger Spiller, Robert Berlin, Richard Swain, Robert Epstein, Lee Dowdy, James Schneider, Jacob Kip, Robert Baumann, Douglas V. Johnson, John Bonin, Henry Gole, Peter Maslowski, and Richard Sommers. I must also thank Drs. Matthew Moten, William Skelton, Edward Coffman, Samuel Watson, Carol Reardon, and Brian Linn, whose excellent histories about the antebellum army continue to motivate

ACKNOWLEDGMENTS

me. Dr. Richard Sommers must receive extra thanks for his continued assistance in my research, along with David Keogh of the U.S. Army's Military History Institute. I also owe a great debt to the staff of the U.S. Army War College and to archivists and administrators of the MHI, as well as the U.S. Military Academy's Special Collection, the National Archives, the Library of Congress, the Historical Society of Pennsylvania, the New York State Library, and the University of North Carolina Library. To Stuart Java, who formatted my maps and figures—thank you. I wish to thank Maj. Sean Wyatt for his assistance and encouragement and my brother, Paul Hope, for his constant support. Finally, I here express my deepest affection and appreciation for my children, Emma and Alec Hope, and my wonderful and patient wife, Karen, without whom none of this would be possible.

A SCIENTIFIC WAY OF WAR

Introduction

In spring 1864 the Army of Northern Virginia, Robert Lee commanding, and George Meade's Army of the Potomac, Ulysses Grant accompanying, bit into each other savagely and pushed and tugged and "chewed and choked" from the Wilderness to Cold Harbor.[1] These armies locked in near continuous contact and in frequent combat tried to destroy one another. Union attempts failed to push Lee's army out of its trenches and sever it from its base of supply. Lee's attempts were frustrated each time the Federals changed their base to a new supply point along the Chesapeake, allowing them to maneuver upon Lee's flank without exposing their line of communications to Washington. In June Lee worried that Grant was switching to a base upon the James. "If he gets there," Lee reportedly said, "it will become a siege and then it will be a mere question of time."[2] The Confederate commander understood that a siege at Richmond-Petersburg could trap his army and deny him opportunity to transfer his base of operations to either the Shenandoah Valley or farther south. Defeat would follow. Gen. William T. Sherman knew as much: "Let Lee hold onto Richmond, and we will destroy his country. . . . Let him stick to his parapets and he will perish."[3]

These generals were becoming masters of the "military science" taught to them at the U.S. Military Academy. They had learned there a theory that defined war as something more than a meeting of armies in open battle, decided by guts, glory, and martial genius.

The reality of three years of civil war, during which destruction of armies eluded both sides, confirmed the theory. So long as an army logistically prepared, made judicious use of fortifications and terrain, and maintained communications with a secure base of operations from which fresh troops, ordnance, and ammunition could come, it could not be destroyed and could be reconstituted to fight another day. Indecisive battle was making generals appreciate the need for strategy, and that strategy was a learned science based upon a sound theory of war.

The Union affirmed its faith in military science when Lee's army eroded during the siege of Petersburg and when Sherman's swath of destruction "round by Georgia and the Carolinas" denied the Confederates any base of operations in the east.[4] Lee later claimed that the collective talent of his generals was insufficient to prevent Southern military might from being "gradually worn down by the combined agencies of numbers, steam-power, railroads, mechanism, and all the resources of physical science . . . these new adjuncts to the science of war."[5] But to Grant and Sherman, and to the West Point graduates who were the commanders and staffs of the field armies, geographic departments, and Washington staff bureaus, victory came not from just numbers or technology but from the proper application of military science itself.

Military science came from France to West Point, New York, twenty years before the establishment of the military academy there in 1802. It was cultivated at the academy following the War of 1812 and became doctrine.[6] Over time, as graduates filled the majority of officer billets in the army, military science became a shared paradigm. The Civil War extended the paradigm, creating an enduring "way of war" favoring logistical, transportation, engineering, artillery, and staff preparedness. The roots of this perception of war, however, were lost in the late twentieth century as powerful revisions heaped layers of criticism upon and effectively buried the intellectual world of the antebellum army. The aim of this book is to challenge these revisions by revealing what constituted nineteenth-century military science, why Americans accepted it as the dominant paradigm, and how it generated an educated understanding of war.

West Point military science was a product of the European Enlightenment, a phenomenon that touched military affairs as deeply as it did civil society. Its genesis was in artillery and fortifications developments in sixteenth-century Italy. What distinguished these developments from their Renaissance antecedents was not technology, as some have argued, but modern mathematics.[7] From Italy, mathematical applications in matters of fortifications and artillery spread to northwestern Europe and began to shape both siege and open warfare. War became the work of large professional armies, with educated leadership, and resembled grand duels of artillery and firearms involving elaborate permanent and field fortifications. The proliferation of firearms and light cannon led to use of firing tables and calculations of troop frontages and ranges. Accurate blueprints replaced scale models for designing fortifications, the art of military map making thrived, and quartermasters and encampment sergeants-major made calculations using logistical, march, and encampment tables. Indeed war during the Enlightenment era steadily became a matter of applied mathematics. These advancements required knowledge of algebra, geometry, trigonometry, logarithms, fluxions (calculus), and were enhanced by use of the decimal system.

Mathematics became foundational to the curricula of military academies that spread throughout Europe in this period. The first of these were private institutions created in Italy in the sixteenth century. Louis XIV established the first state-run school (for artillery) in Douai in 1697, after which such schools flourished. By 1720, France maintained five national artillery academies. The French École du Corps Royale du Genie at Mezieres (1748) and the British equivalent, the Royal Artillery Depot at Woolwich (1741), were for officers bound for artillery and engineer service. The French École Militaire (1751) and the Austrian Military Academy at Wiener-Neustadt (1751) taught infantry and cavalry officers, as did the Royal Academy at Sandhurst (1802). France's École Polytechnique became a military engineering school under Napoleon in 1804 and served subsequently as a model for reforms at the U.S. Military Academy. A common feature of all of these Enlightenment era institutions was the instruction of mathematics applied to war.

Perhaps the most influential practitioner of the new mathematics of war was Louis XIV's chief engineer, Sébastien le Prestre de Vauban. During his fifty-two-year career (1655–1707), Vauban built and improved more than 160 fortresses, perfecting geometrical designs emulated for two centuries. In fifty sieges, Vauban pioneered a method of offensive siege craft—using artillery batteries in an elaborate series of lines parallel to the fortress walls—which could guarantee the capture of almost any fortress. Vauban also perfected a system of integrated fortified frontiers that protected France from invasion and that served as bases of operation from which French troops could invade neighboring territories, a system he called *ceinture de fer* (iron belt). In his innovations, Vauban used advanced mathematics (trigonometry, analytical geometry, and spherical projections) to compensate for variances of terrain in constructing fortifications, in commanding sieges, and in planning his *ceinture de fer*. Vauban's ideas profoundly influenced a group of Enlightenment writers known as *les Lumières*, who were driven by a notion of the perfectibility of war. During the late eighteenth and early nineteenth centuries, these writers tried to determine the invariable principles that would unlock the secret of open warfare, using Vauban's ideas of fortifications, siege craft, and military frontiers.[8] In this, mathematics, especially geometry, became their principal tool, and *les Lumières* started to refer to many aspects of military affairs as "sciences."

Today *science* is understood within the dichotomy of art and science, and *a science* refers to a very specific body of knowledge concerned with the physical world. In the eighteenth and nineteenth centuries, a *science* was any branch of learning that had a theoretical construct containing principles or rules that explained empirical evidence and that permitted consistent transfer of knowledge.[9]

Astronomy and natural philosophy (physics, chemistry, physiology, and botany) were the common sciences, but by the late eighteenth century, politics and military affairs became legitimate sciences as well, with Enlightenment writers cataloging their governing principles. Sciences were distinguishable from *conscience* (philosophy), which considered moral reason without empirical evidence, enduring principles, or mathematical underpinning. Sciences could also be components of an *art*—considered skill at achieving something. The science of physics, for instance, could enhance an architect's talent in the art of structural design.[10]

Enlightenment military science was transferable knowledge of war. Its limited use in the early eighteenth century described the work of artificers, artillerists, and engineers. By midcentury, knowledge regarding the raising and equipping of armies (elements now referred to as mobilization), quartermaster and ordnance functions, the conduct of campaign marches, encampments, and orders of battle—everything that teachers could reduce to general principle and mathematical calculation and pass on by instruction—came also to be known as military science. After 1800 it included the "science of movements" or, as it was later referred to at West Point, "the science of war" or "strategy."[11] Knowledge in any of these "sciences" could enhance a general's talent in the art of war.

For purposes of clarity, this work defines military science as it was perceived in the antebellum military academy as including the science of war (strategic movements or military campaigning) as well as three other specific branches that relied on mathematics: artillery applications (including ordnance functions), fortifications and engineering, and the organizing, supplying, and encampment of armies (logistics and administration).[12]

Nineteenth-century military science also required knowledge of topography. West Point cadets learned the Vaubanian precept that all military operations had to conform to terrain because they were "controlled by the topographical features of the seat of war."[13] Fortifications, the use of artillery, and battle formations had to be adapted to local conditions. The limits for movement of men and material were determined by terrain features and constraints of foot, horse, wind, and steam power. A land's capacity to feed an army was also an important consideration in campaigning. Appreciation for these topographical aspects allowed nineteenth-century officers to distinguish between strategy and tactics, the definitions of which deserve consideration.

Nineteenth-century strategy was a cognitive deliberation pertaining to the establishment of national military frontiers and the identification of "theaters of war," "objectives," "bases and lines of operation," and included the "science of movements." Strategy involved the selection of important topographical points for the construction of fortifications on land and coastal frontiers, the designation of "strategic points" for the mustering, organizing, and supplying of armies, and the planned movement of these forces along strategic lines to concentrate them at selected topographical points on or forward of the frontiers. The designation of a theater of war and strategic points and lines came together in what was referred to as a campaign plan, and strategy pertained to military movements made during the campaign out of range of an enemy's weapons.

Military tactics, on the other hand, were the artful employment of forces—concentrated at a strategic point—within cannon range of the enemy.[14] Tactics involved calculation of weapons' effects on specific types of terrain, the determination of the best angle for soldiers to advance upon an enemy in parallel, oblique, or perpendicular lines, and the use of engineering works to facilitate or impede the effects of fire or movement. Understanding weapons effects was necessary to maximize the fire of muskets and cannon, and this led to the linear formations that characterized eighteenth- and nineteenth-century tactics.

An essential feature of both strategy and tactics was the manipulation of physical space for military purposes. At the tactical level, this meant constructing field or permanent fortifications to dominate terrain by concentrating weapons fire or improving roads or bridges to best move and concentrate masses of soldiers. At the strategic level, military science involved the construction of depots and barracks and the creation of integrated fortified frontiers and road, water, and rail transportation networks. The governing principle at both levels, in all situations, was thus: the side that could best utilize natural and "improved" terrain to bring the greatest mass of soldiers and weapons fire upon a given geographic point of tactical or strategic relevance would probably prevail. In 1805 Jomini described this fundamental principle as follows: "The grand object is then first to know the point upon which this concentrated effort should be made: secondly, so to make it, as that the adversary shall not be prepared to meet it by an effort equally concentrated on that point. This may be said to be the outline of military science."[15] The principle was explanatory, giving military thinkers "a system founded upon truth" to explain the successes of Frederick the Great and Napoleon Bonaparte so that their victories were not considered "the effects of magic." However, despite what some modern critics believe, military science did not equate to predictability. Enlightenment thinkers knew that war, like life in general, consisted of factors other than mathematics, and was subject to human agency and contingency.

Military science reflected broad social trends, particularly the emergence of an intellectual milieu named by French philosopher Blaise Pascal (1623–62) as *l'esprit géométrique*. Early Enlightenment thinkers considered geometry key in understanding the workings of the universe. God was the great "Geometer," and geometric form dominated architectural and artistic design. Pascal believed that geometry alone explained natural order but not matters of conscience, and he therefore divided human thought into two categories, one mathematical, the other intuitive. The *mathematical mind* viewed natural phenomena as riddles to be solved using arithmetic and early scientific method.[16] The *intuitive mind* was innate and given to artistic expression and a quality known as genius. Pascal's dichotomy might seem simplistic today, but it expressed a duality used subsequently by Enlightenment writers (and West Point professors) to distinguish military science based upon geometry and empirical reasoning learned in academies from more intuitive military talent or genius honed by experience.

In treatises written by *les Lumières* and in subsequent American works, the authors had to deal with the question of learned military science in relation to the vague but indisputably important quality of genius. Most followed Pascal's model and divided military knowledge into two distinct parts, one scientific and one intuitive (referred to by many as the "sublime" and commonly called *coup d'œil*).[17] The writers were distinguishable in the emphasis each placed upon one or the other of these parts. One school, led by English and later Prussian writers, advocated the overall importance of personal talent or genius. The other school, championed by French and some German thinkers and accepted by Americans in the antebellum period, held that talent and genius were important when applying tactical principles, but that strategy required military science, learned through education and experience in service of the state.[18] The best commanders—personified by Frederick and Napoleon—were men whose natural genius was refined by serious and formal study. The emphasis on education also facilitated the development of military staffs. Indeed, military science elevated the role of the staff in the preparation and the conduct of war. The place of the academy in creating strategic thought and staff competency, in contrast to mere reliance upon natural individual genius, was important in this era and is therefore an important aspect of this book.

Differences in interpretations of the role of genius and science in war define the current historiography of antebellum military affairs. A late twentieth-century liberal arts perspective now prevails, dismissing any sort of Military Enlightenment corresponding to the Enlightenment in civil society and refusing to acknowledge the existence of a "complete system" of military education or the possibility of any science to warfare.[19] Instead, the consensus supports an appreciation for the metaphysics of Prussian theorist Carl von Clausewitz, with its essential assumption that all war, and therefore strategy, is characterized by incessant tensions between reason and passion and is governed by chance, expediency, and genius.[20] Enlightenment ideas of learned military talent have faded away in this revision. The most renowned military theorist of the time, Swiss-French general and author Antoine Henri Jomini, whose works are exemplary of Enlightenment clarity of reasoning, has been cast in the role of *bête noire* opposite Clausewitz. Nineteenth-century military education is now misjudged as having been too mechanistic, narrowly technical, restrictively "Jominian," and prohibitive to the professionalization of the officer corps of the U.S. Army.[21]

This book deviates from consensus interpretations that antebellum military education was entirely a throwback to the eighteenth century and that soldiers were merely tradesmen in American society. It demonstrates the contemporary relevance of the doctrine of military science, its political acceptability in the War Department and Congress, and its importance to emerging professionalism. The chapters move beyond the cataloguing of separate ideas regarding artillery, fortifications, logistics, and strategy and explain instead the coherency of this "complete system," placing the doctrine into the context of a national military policy and a changing social environment. The work shows that antebellum military science constituted a pervasive mind-set derived from doctrine that complemented evolving political and social trends.

The central focus of this work is the intellectual element of West Point teaching that some might refer to as a "way of war." Since its publication in 1973, the enduring influence of Russell Weigley's *The American Way of War* requires that one take care with all subsequent usage. Weigley proposed that a way of war was a choice of one of two types of military policy. He concluded that Americans are predisposed to one type—annihilation—over another—attrition.[22] Doctrine and organizational military culture had no place in Weigley's dichotomy. This has provoked criticism, most notably by historian Brian McAllister Linn, who has observed that Weigley's idea about policy choices existing independent of organizational culture is problematic. Linn suggests a broader definition of this thing called a way of war, involving the peacetime preparation of an army, including decades of consistent intellectual development, and inculcation in a particular doctrine of war that is evident in the instruction and transmission of ideas.[23] Linn equates his way of war to what Gen. John Galvin described as a "comfortable vision of war . . . that fits our plans, our assumptions, our hopes, and our preconceived ideas." Samuel Huntington has referred to this notion as military mind. In its broadest sense it is a dominating military thought, the *mentalitée* referred to by historians Lévy-Buhl and Marc Bloch.[24] One might also consider a way of war as the "paradigm" espoused by Thomas Kuhn in *The Structure of Scientific Revolutions*.[25] This work distinguishes between military doctrine (that which is taught), a paradigm (a common understanding of war derived from instruction and inculcation in doctrine), and a way of war (the common application of military doctrine based on a shared paradigm). The intellectual component of peacetime preparations, evident in the continuity of military doctrine and a paradigm of war, is the center point of this work.

I attempt here to demonstrate that the doctrine inculcated at West Point in the antebellum period, called military science, containing an enduring and coherent military theory, was the foundation for broader American military thought that West Pointers applied in the Civil War. The doctrine came not from any particular strategy or ideas of policy choices but from a prevailing—perhaps obsessive—intellectual movement that sought mathematical and scientific explanation for the phenomenon of war. The movement originated in the Enlightenment, and during the late Jeffersonian era was embraced by officers such as Winfield Scott, Robert E. Lee, Albert Sydney Johnston, Joseph Totten, and West Point professors such as Dennis Hart Mahan, men who shared ideals regarding the governing role of science in human affairs. It was taught to, and maintained thereafter by, Jacksonian officers such as Irwin McDowell, George

INTRODUCTION

McClellan, Pierre Beauregard, Ambrose Burnside, Joseph Hooker, Braxton Bragg, Henry Halleck, Thomas Jackson, John Bell Hood, George Meade, John Pemberton, Ulysses Grant, William Sherman, and hundreds of others as a "comfortable vision" about war. At the same time, this educated perception of war met the requirements of the system of national defense and satisfied the enduring Moderate Whig tradition in American politics. The regular army maintained the intellectual predisposition toward mathematics and science so well that it continued to be taught at West Point after the war and long after civil society in America had displaced Enlightenment and Jeffersonian ideals with Jacksonian Romanticism and late nineteenth-century Realism.

I have divided this work into eight chapters organized chronologically to track the evolution of the doctrine of military science from its transplantation from Europe during the Revolution to its application in the Civil War. Chapter 1 covers the period of the early republic from the Revolution to 1812. It describes how military science, although used in the Revolution, fell victim to Radical Whiggism after 1783 and only began to take root after the establishment of a corps of engineers and artillery in 1794 and the U.S. Military Academy in 1802. Military science remained limited and little known until British amphibious incursions during the War of 1812 convinced the government of the need for military reform, the subject of chapter 2. The reforms included the adoption of French Napoleonic doctrine for West Point and a defense policy involving a network of permanent coastal fortifications.[26] These reforms remained intact during the antebellum period and, together with a corollary effort of "Internal Improvements," saw Congress commit substantial resources to build an American *ceinture de fer* containing hundreds of land and maritime fortifications, roads, and waterways and to underwrite the military academy that made such efforts possible. Between 1820 and 1860, this policy framed thinking in the War Department and absorbed most defense expenditures. It also substantiated the need for a national military bureaucracy and a small corps of educated regular officers distinguishable from their amateur counterparts of the states' militias by their military knowledge and by their sense of duty in federal service.[27]

Chapter 3 examines what West Point taught these regular officers during their cadetship. It details the "complete system" of thinking offered by French military science dedicated to state-on-state warfare, or *la grande guerre*. This doctrine provided the foundations for professionalism, developing habits of thought and conduct that subordinated notions of individual honor and state identification to a sense of federal duty. It also reinforced a Jeffersonian ideal of egalitarianism. Military science remained accessible to any youth possessed of intellect and determination to learn. The curriculum was unique in that it dedicated two years of study to mathematics and two years to the study of physical science and engineering. Equally unusual was the requirement that every cadet become competent in all tactical branches, mastering the basics of infantry, artillery, engineer, and eventually cavalry tactics. The cadets received no branch affiliations until graduation. The curriculum also provided every cadet with a body of knowledge of ordnance and logistics, military engineering and fortifications, and the science of military movements—or strategy. Cadets were taught that military decision-making was a matter not of individual intuition or experience but of enduring principles founded upon historical practice and mathematics. This curriculum challenged aristocratic notions that effective command was the exclusive domain of officers from a privileged social group. It also challenged ideas held in some quarters that amateurs possessed of natural talent would suffice in war. Cadets graduated with specific knowledge that might develop good commanders but which surely produced good staff officers.

Expansion westward forced the U.S. Army to augment its staff capacity, and between one-quarter and one-third of the regular officer corps were seconded to staff branches. Chapter 4 explains this and also how expansion affected the West Point curriculum. The faculty added advanced civil engineering and steam technology in order to prepare officers attached in temporary service to the engineer, topographical, ordnance, or quartermaster staffs to construct roads, bridges, and garrisons in the west and to build canals, harbors, railways, depots, and arsenals in the east. Officers performing these duties were federal agents working alongside state officials and private industry to complete projects of considerable scope—sectional and eventually national in their design.

Jacksonian politics and social changes also brought modifications to the academy, the subject of chapter 5. I reveal here how Professor Mahan, head of the engineering department and the man charged with teaching strategy, did away with European texts and replaced them with his own synthesis of strategic theory. Mahan insisted that all cadets acquire basic knowledge of permanent fortifications based on Vauban's designs, refusing to teach sophisticated and nuanced European variations. In contrast to European armies, Mahan felt that all officers, regular and militia, regardless of branch of employment also needed basic knowledge of field fortifications, march discipline, military encampments, reconnaissance, and logistics. He set out to deliver this knowledge in the classroom and in published texts.

West Pointers successfully applied Mahan's strategic combinations and the logistics, ordnance, siege craft, and reconnaissance aspects of military science in the Mexican War, covered in chapter 6. After that conflict, military science was refined and diffused into the U. S. militia through a series of publications. Concurrently, thinking about strategy and fortifications was elevated within the regular army when professors began to teach advanced military studies to the academy instructors in the Napoleon Club and in Mahan's graduate-level courses for engineer officers teaching at West Point. Chapter 7 examines further dissemination of military science in the 1850s through published works and the commencement of a five-year program of study wherein cadets received instruction in military history and strategy. Strategic thinking was not purely theoretical. Army officers continued to perform logistics, survey, and construction functions that were truly continental in scope and therefore strategic in aspect, and the experience gained in this work mattered during the Civil War. Defense policy reviews, the 1856 Delafield Commission, and the reappraisal of West Point that occurred with the 1860 Davis Commission affirmed its curriculum as America's "comfortable vision" of war.

Chapter 8 suggests that military science affected the Civil War. It introduces a hypothesis that the Civil War saw a broad diffusion of military science by the hundreds of West Point graduates who became the commanders and general staffs on both sides. Throughout the war, these officers applied knowledge of mobilization, logistics, ordnance, fortifications, siege craft, reconnaissance, marches, field engineering, encampments, and campaign design. The events of 1864–65, including Grant's crossing of the James, the siege of Petersburg, and Sherman's march through the South's interior bases of operations, demonstrated an epitome of applied military science.

I draw my conclusions in chapter 8 from public documents and commentaries; personal memoirs; the biographical records of academy graduates; employment records in annual *Army Registers* from 1820 until 1861; and Civil War records of how armies organized, moved, and fought. Some of the results are set down in tabular form and provide telling information regarding the extent to which West Point graduates were employed on fortifications engineering, internal improvement, general staff, and instructional duties during the antebellum period and how these same graduates were employed during the Civil War. The tables are derived from eighty-eight pages of detailed biographical notes on 2,046 graduates, available to the reader in my original dissertation via the internet.[28] The data challenges the well-established notion that West Point graduates never learned or experienced command above the company level and were unprepared for service in large army units. Specifically I challenge the assertion that successful generalship in the war was simply a result of natural individual talent and not connected to previous military experience or doctrine. I accept the criticality of personality in war, especially in battle. However, my research and substantial combat experience have convinced me that paradigms held before conflicts commence have great impact on both individual and collective performances in war. I also challenge the notion of inevitability of outcome. Despite what Lee and Lost Cause advocates suggest, the Union victories in 1865 were not a mere matter of overwhelming numbers. Contingency remained in play until very late in the war. The education and experience of commanders was influential in this dynamic, but not decisive.

Nor was West Point military science the only choice for a theory of war in the antebellum United States. Multiple sources of education and training, sectional differences regarding military service, the dispersion of the army across the continent, and widely diverse experiences of individuals and units presented alternative theories and precluded universal agreement on any single perspective of war for both regular and volunteer alike. One alternative was the disbandment of regulars and the academy and reversion to total reliance upon state-controlled citizen militias, instituting an American equivalent to the French revolutionary *levée en mass*. Innate martial genius would be found among these armed citizen forces. Alternately, some state and federal politicians favored keeping the regular army but forcing its adoption of an irregular style of warfare—*la petite guerre*—that had long roots in North America, with an Indian-fighting and not a European focus.[29] Indeed Americans had practiced *la petite guerre* since first landing, and it was an integral part of the America militia and regular army legacies. Other Americans preferred converting to the English or continental military system, with their larger standing armies officered by men appointed from the social elite and emphasis on individual personality, experience, and talent over formal study.[30] Despite these alternatives, the staff in Washington and the professors at West Point throughout the antebellum period accepted the small size of the regular army and the reliance upon mobilized state volunteers, and ignored anything but their French-influenced and educated approach to war. Their minds never changed, and West Point maintained an impressive continuity of instructional content for more than fifty years. Professor Mahan spoke of the alternatives and preferences in 1847 when he stated:

> The system of tactics in use in our service are those of the French; ... there is really more affinity between the military aptitude of the American and French soldier, than between the former and the English; ... French systems are the results of a broader platform of experience, submitted to the careful analysis of a body of officers, who, for science and skilled combined, stand unrivalled; whereas the English owes more to individual than the general talent; and therefore is more liable to the defects of individual pride of opinion.[31]

The purpose of this work is to show that this "system of tactics" taught at West Point, based on a theory of war as science, was maintained deliberately as the dominant antebellum military doctrine, which, by the end of the Civil War, became foundational in American thought. This paradigm maintained faith not in natural individual genius but in collective acceptance of an educated, and therefore scientific, way of war.

I

Colonial and Early National Military Science

When fighting began in earnest during the American Revolution, military and political leaders realized that the methods of *la petite guerre* were not sufficient to beat the might of England, and they sought to acquire European engineering, administration, and artillery expertise to instruct the Continental Army. However, attempts to institutionalize military science ceased with the end of hostilities. Lack of clear threat and changing political and economic circumstances, together with Whig traditions and an accepted narrative of the militia's effectiveness, beclouded the need for professional military forces or education. Appreciation for military science stagnated. Change came with the renewed threats of war after 1790, when a system of coastal fortifications was attempted. It was not until the establishment of the military academy in 1802, however, that military science began to influence thinking in the small regular army and the states' militias alike.

Colonial Military Science

Minutemen at Lexington-Concord in April 1775 needed no military science. However, militiamen around Boston Harbor just weeks later began to suffer from a lack of artillery and engineering expertise and from want of a system of supply and administration. Upon assumption of command of the army at Cambridge in July, Gen. George Washington advised the Continental Congress to correct these deficiencies.¹ American leaders at this time were facing problems inherent to their colonial heritage.

The plantations at Jamestown and Plymouth had been established before the emergence of *l'esprit géométrique* and before Vauban and les Lumières attempted to perfect state-run warfare. For their first eighty years of existence, these colonies and those that followed had maintained limited military establishments. While seventeenth- and eighteenth-century Europeans lived under constant threat of invasion from large standing armies never more than a few marches away, European threats to the English colonists remained occasional and distant. Moreover, while threat of invasion in Europe produced sophisticated militarized frontiers and spurred advancements in military science, the lack of such a threat in North America gave no equivalent impetus for military education in the colonies.

However, the first plantations were sufficiently militarized to produce a legacy. All male colonists brought armor and weapons over the ocean, which they then carried constantly and hung on the walls of their makeshift homes as they slept. Military practices were part of everyday life. Edward Winslow described the exercise of arms in front of Native guests during the first Thanksgiving Day in 1620. At all times the colonists relied on armed guards, wooden fortifications, and defensive musters when threats emerged.² This is not to suggest that the colonial plantations were military exercises. The carriage of arms and the performance of drill during village muster were defensive measures to counter the threat of raids by European foes and Indians. But to expand their territories colonists preferred the planted seed to "Swords, Rapiers, and . . . other piercing weapons."³ In the Christian idea of their day, appropriations of new farming and grazing land and subjugation of the Indians, even by the use of armed force, were natural and correct and did not warrant the name *war*. That was reserved for monarchs, the sovereign being "the real author of war" and colonial subjects being "only instruments in his hands."⁴ Without sovereignty of their own, "war" was not the English colonists' to make. Forced subjugation of the Indian was viewed as a constabulary measure to suppress rebellion and not a martial endeavor. Besides, the Indian possessed no artillery, and elementary military knowledge sufficed to deal with this threat. It was not until the late seventeenth century, when imperial wars became incessant, that colonists felt the need for even a small measure of European military science.

For much of this period there was no English army presence in North America and no corresponding need for military schools. Education for the colonial militia remained limited; they trained under veterans of wars in the "Low Countries," the English Civil War, or the suppression of rebellion in Ireland. They used Richard Elton's *The Compleat Body of the Art Military* (1650) and William Barriffe's *Military Discipline; or, The Young Artilleryman* (1635), drill manuals that also formed the basis for numerous laws written to govern the training of colonial militias until 1750.[5] Militia officers also acquired knowledge from personal libraries and in the eighteenth century from service with British regulars periodically deployed to the colonies. However, most military books came from England, where they favored revised Greek and Roman texts.[6] English-language works that incorporated elements of modern military science only became available to colonial militia officers in the mid-eighteenth century and included Benjamin Robins's *New Principles of Gunnery* (1742), Humphrey Bland's *A Treatise of Military Discipline* (1743), John Muller's *A Treatise Containing the Elementary Part of Fortification* (1766), and Roger Stevenson's *Military Instructions for Officers Detached in the Field: Containing, a Scheme for Forming a Corps of a Partisan* (1775). However, because there were no schools requiring the reading of such literature, these works were used only for personal study or in scientific societies.[7] The evolution of military thought and education in colonial America therefore differed substantially from that in continental Europe, and the colonists' military knowledge remained rudimentary until the Revolution.[8]

The informal nature of military instruction did not mean that colonists were totally ignorant of emerging aspects of military science. Fortifications were a part of colonial life, and militias understood how to mount and employ single cannon. But it took a series of imperial wars to introduce more advanced elements of military science. European powers contested colonial frontiers during King William's War (1689–97), Queen Anne's War (1702–13), King George's War (1744–48), and the French and Indian War (1754–63). These involved raids and expeditions and the construction and sieges of scattered forts, conducted in the deliberate manner of period European warfare but on a much smaller and more dispersed scale. In each, French and British engineers introduced European military techniques and permanently altered the colonial landscape by building durable stone fortifications and military roads and bridges. However, it was regular French and British engineers and artillerymen, trained at Mezieres and the Royal Artillery Depot at Woolwich, who provided this skill. No indigenous colonial expertise emerged during this period, and at the commencement of the Revolution there was only one commissioned American engineer officer and no school teaching military architecture or artillery techniques.[9]

Americans were also reluctant to accept military education because of its association with the maintenance of standing armies, defined as units of soldiers permanently established and maintained as part of society. The idea of a substantial standing army was problematic to the colonial psyche, which perpetuated Radical Whig sentiments that vigorously opposed permanent military establishments, preferring instead the Elizabethan system of universal military service.[10] None of the early colonies maintained permanent bodies of troops. The wooden defenses of the first settlements were instead manned by a "watch" consisting of a number of male citizens who stood guard at village entrances and atop the fortified blockhouse that became a feature in most new plantations. The watch came from the "trainband" of the settlement, consisting of every able-bodied male, regardless of color or religious affiliation.[11] The trainbands were the original militia, formed into companies within each town, officered by elected men. Their mandate was purely protective, to watch and to defend the commons in the event of an attack. They were a direct transplantation of the Tudor militia tradition. Their duties were mostly sedentary, but they sometimes performed constabulary tasks outside of the plantation. Military knowledge in the trainbands never rose above the elementary practice of company musket, pike, and sword drill and the use of a cannon.[12]

Most colonies wrote this ideal of universal military service into early colonial charters. However, along the populated east coast, once the immediate threat of Indian war diminished in the mid-eighteenth century, militia training atrophied into token practice, continuing as a social venue more than a military function.[13] To a degree, volunteer companies and regiments compensated for this neglect.

By the middle of the seventeenth century, affluent colonists who wished to secure commissions without election began to raise and equip volunteer companies, providing an offensive capability for ventures beyond the county limits. Together with trainbands these two parts of the military establishment precluded need for a standing army. By the middle of the eighteenth century, the companies in some colonies combined in emergencies to form provincial regiments, composed of sufficient manpower and equipage to mount long expeditions. These forces provided auxiliary military capability to the British regulars periodically garrisoned in the colonies during conflicts against France and Spain. The volunteer provincial regiments of the French and Indian War became indispensable and were prototypes for Washington's Continentals during the Revolution.

Volunteer companies also served as associations for transmission of knowledge, including subjects of military science such as casting cannon, storing gunpowder, building local fortifications, and mounting artillery pieces. Perhaps more important, volunteer companies served to pass on knowledge of *la petite guerre*. Historian John Grenier suggests that this colonial practice became the first "American way of war," distinct in its preference for attacking Indian noncombatants and its tendency toward extremes of violence.

Grenier traces the evolution of this way of war from 1622 to 1815, highlighting its predilection for punitive patrolling tactics used to destroy enemy crops, food stores, and villages and for scalping and extirpative warfare. Generation after generation of colonists learned such techniques, forming a paradigm of warfare that, according to Grenier, helped construct an American identity characterized by admiration of the American frontiersman as a killer.[14]

While somewhat problematic in its insistence that small unit "ranging" tactics constituted the entirety of early American military thought, Grenier's interpretation is nonetheless valuable, as it allows us to see that adoption of European military science for *la grand guerre* was not inevitable and that alternatives were always readily at hand. During the Revolution Washington was acutely conscious of this. He found Gen. Charles Lee quick to argue for "partisan" warfare, an alternative to the conventional approach that Washington preferred.[15] Successful irregular operations in the Mohawk and Ohio valleys and in the Carolinas reinforced the argument. Indeed, widespread use of Patriot militias to neutralize Tory elements and to limit the freedom of movement of British regulars and Loyalist militia was creating faith in the militia's ability in *la petite guerre*, laying the foundation of a narrative of the innate military power of militiamen fighting as partisans or ranging against Indians.[16] Post-revolutionary campaigns by Generals Anthony Wayne and Andrew Jackson would add to the narrative and provide a clear alternative to the European military science that Washington was desperate for during and after the Revolution.

Facing the prospects of a long siege at Boston in 1775, Washington felt an acute need to break away from the militia mind-set and to secure proper quartering, cannon, and skill in siege entrenchments. In July Congress responded in part by approving foundries and depots for most colonies, allowing the casting of cannon and the stockpiling of supplies. They also approved the establishment of an artillery cantonment at Pluckemin, New Jersey, where Chief of Artillery Henry Knox could teach and experiment. However, the need for engineers was harder to address, made even more so when colonies, desperate to fortify areas open to British invasion, solicited any officers who possessed knowledge of civil engineering to take off the uniform and come under their employ.[17]

In December Congress resolved to secure the services of trained engineers from abroad and in April 1776 sent an agent to France for this purpose. France was an obvious choice: not only was it a traditional enemy of England but it was also the center of scientific education, with the École du Corps Royale du Genie producing Europe's best military engineers. American agent Silas Deane recruited several Frenchmen who arrived in Philadelphia in spring 1777, the first of a steady trickle of individually sponsored European engineers who came to America over the next three years, the most famous being Gilbert du Motier, marquis de Lafayette. At the same time, Benjamin Franklin—in Paris to promote a treaty of alliance with France—gained King Louis's approval for the dispatch of four Royal French Army engineers to support the American war effort. Led by Louis Lebègue Duportail, these officers presented themselves to Congress in July 1777 and began their work later that month.

At any given time, the Continental army employed a dozen foreign engineers, while Richard Gridley and Rufus Putnam served as the only American engineers.[18] Washington also received congressional permission to employ Robert Erskine as the army's first geographer, and Erskine sought out assistants with "Mathematical genius . . . acquainted with the principles of Geometry" to survey roads and to prepare maps.[19] Congress also approved the employment of Prussian Friedrich Wilhelm von Steuben as inspector general of the Continental Army. Von Steuben, seeing the want of military science, enforced common standards in basic tactical evolutions and weapons drill, and instituted a system of military administration and field discipline, including orders of battle and rules for marches and encampments. His 1779 *Regulations for the Order and Discipline of the Troops of the United States*, known as the *Blue Book*, served the Continental Army and the U.S. Army for over forty years.[20]

Washington was not so quick to gain congressional support to create an American corps of engineers. He wanted three companies of sappers and miners able to organize and employ fatigue parties in the preparation of fortifications. Although he made his proposal in 1776, it only received political support in 1778, with the unit forming in 1780. The establishment of this organization marked recognition by military and political leaders of the need for indigenous capability in field fortifications and siege craft, to compensate for "deficiency in the practice of manoeuvres" by the Continental Army.[21]

Training and educating the new engineer corps was a special undertaking. Although instructors in mathematics were attached to each company, it was fieldwork itself under experienced French officers that allowed the rank and file to learn their trade.[22] There was much opportunity for this. Field and permanent fortifications and sieges became commonplace in the American Revolution. Fighting at Boston, Quebec, Fort Washington, New York, Ticonderoga, Bennington, Freeman Farm, and Saratoga involved fortifications, and the bases of operations at Valley Forge on the Delaware River and throughout the Highlands on the Hudson River were heavily fortified. In the South, sieges at Savannah and Charleston preceded the decisive siege at Yorktown in 1781, all conducted using standard Vauban methods involving siege parallels and the use of grand batteries.

Military science also emerged in army ordnance and logistics. The Continental Army acquired depots at West Point, New York, and at Carlisle, Pennsylvania. The Continental Congress also authorized the establishment of a Quartermaster Department, a Commissariat, a Clothing Department, an Ordnance Department, and a Hospital Department. However, there existed no militia staff system to coordinate logistics, ordnance, and other duties, and knowledge was confined to those few militia officers who had read books or had experience in service with British regulars. Humphrey Bland's *A Treatise of Military Discipline* was helpful, but Quartermaster Generals Thomas Pickering and Nathanael Green and Chief of Artillery

Henry Knox also learned their craft from French texts by Marshal de Saxe and Turpin de Crisse.[23] This limited military knowledge was combined with diverse experiences in mercantile capitalism to provide a basic degree of expertise in procuring and distributing supplies through contracted storekeepers, conductors, artificers, and laborers. However, the staff departments were organized and worked according to the talents of the men who led them, and problems of decentralization and lack of standardization plagued the ordnance, supply, and transportation efforts of the army until war's end. Another major problem was lack of continuity. The Quartermaster General's office, for instance, changed hands four times in its first three years, with each incumbent growing weary at the effort required to procure and supply the army and of the complete lack of glory associated with his duties.[24] Congress had shut down most staff departments by July 1785.

When hostilities ended in 1783, most officers of the Continental Army understood the importance of military science, especially of engineering and artillery. Before he left for France that year, the chief French engineer, General Duportail, submitted recommendations to Congress for the establishment of a permanent combined corps of engineers and artillerists, believing that the United States could not afford the European luxury of specialization between the two functions and showing great understanding of the American condition.[25] The vastness of the country and the paucity of experience and technical skill would work against specialization of function for another century. Each state in the union needed persons possessed of multiple skills. The army, if one was allowed in the postrevolutionary United States, was no exception.

Duportail suggested that the new union maintain sufficient numbers of engineer-artillerists to fortify frontiers exposed to renewed British attack and that all frontiers be reconnoitered to find "the most proper places for the Forts and for all the Establishments relating to War." Suspicious of independent states' governments, Duportail recommended a "general plan of defence" containing three distinct frontier regions, each served by a unit of engineer-artillerists under a single senior officer responsible for all fortifications on that frontier, regardless of state boundaries. These officers would report to a director of a federal department of artillery and fortifications, who in turn would report to the "Board of War" or to Congress directly, circumventing anyone other than the highest national authorities. Duportail suggested that the government find and appoint "a Vauban" to carry out the duties of director.[26]

Duportail also proposed the creation of an academy for engineer-artillerists, suggesting a three-year curriculum of mathematics, chemistry, natural philosophy, and drawing. Having made his pitch, he went home to France in spring 1783. One of the remaining French engineers, Maj. Pierre Charles L'Enfant, took up Duportail's efforts. He recommended a federal system of peripheral fortifications, which he suggested was a "Continental undertaking" to secure all frontiers, "not only . . . the inland Boundary but . . . all the Sea Coast and . . . every Harbour and Entrance." This system had to be under federal management if they wanted to meet standards and control expenses.[27]

L'Enfant also promoted education, describing arithmetic, geometry, mechanics, architecture, hydraulics, drawing, and natural philosophy as essential. The duties of American engineer officers would include the surveying, planning, and construction of fortifications, magazines, and arsenals for mobilization and the calculation of peacetime garrison requirements for each post. His statements were remarkably prescient. L'Enfant, like Duportail, was transplanting the notion of Vauban's *ceinture de fer* to America. While not endorsed by Congress, L'Enfant's ideas did resonate with American generals and Nationalist politicians.

Early National Military Science

Five leaders of the Revolution—George Washington, Henry Knox, Alexander Hamilton, James McHenry, and Friedrich von Steuben—appreciated the utility of knowledge of artillery and engineering and supported L'Enfant and Duportail's proposals. Knox had initiated attempts to establish a military academy in 1776 and succeeded in conscripting John Adams to his cause. Adams pushed Congress to accept the formation of a cadre of disabled veterans (the Corps of Invalids), which eventually formed at West Point in 1780 to oversee an arsenal, a library, and a school for propagating scientific military knowledge. Congress abandoned this effort in 1783, although a small garrison of artillery remained at West Point thereafter.

George Washington, grateful for French assistance, wanted to reduce dependency on foreign military engineers by supporting the establishment of an American school for military science.[28] In 1783 he collated most of the ideas of his engineers and generals into "Sentiments upon a Peace Establishment," wherein he stated: "I cannot conclude without repeating the necessity of the proposed Institution, unless we intend to let the Science [of war] become extinct, and to depend entirely upon the Foreigners for their friendly aid."[29]

Washington's "Sentiments" was well constructed and strategically minded. If it had been endorsed, it would have become the first coherent defense policy for the United States, superior to all subsequent attempts until the Third System policy of 1820 (which incorporated most of Washington's ideas). Washington's proposals represented the ideas of Nationalists (and Federalists after the Constitutional Convention) that would eventually become the prevailing perspective of the early nineteenth-century U.S. government. Specifically, Washington suggested that a small regular force protect inland frontiers from the British and the Spanish and "awe the Indians." The force would comprise infantry, artillerists, and engineers with the scientific and technical skills needed in modern war, which Washington knew was becoming too complicated for militiamen and volunteers. He proposed an integrated network of defensive works on the northern and western frontiers, connected by military roads and waterways and garrisoned by 2,631 soldiers— enough for one company of infantry and a detachment of artillery at each of the proposed defensive locations.[30]

Beyond the regular component, Washington wanted a national system of mobilization that incorporated the traditional colonial practices of universal service and selected service for expeditions and frontier duty. The universal service militia needed to be uniformly equipped, exercised, and able to turn out for local defense. The selective service component would comprise every tenth, fifteenth, or twentieth universal serviceman, who would remain on active duty for several years in a "continental militia." Additionally, however, Washington proposed the creation of three great arsenals where "might be deposited, Arms, Ammunition, Field Artillery, and Camp Equipage for thirty thousand Men, Also one hundred heavy Cannon and Mortars, and all the Apparatus of a Siege, with a sufficiency of Ammunition," and academies "for the instruction of the Art Military."[31]

Influenced heavily by French thought, Washington set down the conditions that would later define the U.S. Military Academy at West Point and distinguish it from European practices:

> Provision should be made . . . for instructing a certain number of young Gentlemen in the Theory of the Art of War, particularly in all those branches of service which belong to the Artillery and Engineering Departments. . . . And as this species of knowledge will render them much more accomplished and capable of performing the duties of Officers, even in the Infantry or any other Corps whatsoever, . . . appointments to vacancies in the Established Regiments, ought to be made from the candidates who shall have completed their course of Military Studies and Exercises. . . . The Regiments of Infantry by this means will become in time a nursery from whence a number of Officers for Artillery and Engineering may be drawn on any great or sudden occasion.[32]

Washington's opinion that education was fundamental to all officers was well before its time and showed appreciation for military science. However, lack of knowledge of all its branches was evident in his proposals. Of maritime frontier defenses he confessed ignorance: "Fortifications on the Sea Board . . . is a matter out of my line, and to which I am by no means competent." In the areas of ordnance and logistics, Washington preferred that Congress seek the advice of von Steuben and Alexander Hamilton:

> That some kinds of Military Manufactories and Elaboratories may and ought to be established, will not admit a doubt; but how far we are able at this time . . . I leave those to whom the observations are to be submitted, to determine, as being more competent, to the decision than I can pretend to be.[33]

Congress rejected Washington's proposals, largely because anti-Nationalists led by Massachusetts representative Elbridge Gerry refused to endorse a federal military force whose purpose he believed was to protect New York's fur trade. In 1784 Congress instead authorized a militia levy of seven hundred men to garrison frontier forts for one year and replaced this in 1785 with another levy for three years. These ill-disciplined troops followed settlers moving west on the Ohio River into Kentucky and Indiana. Their inability to stop unauthorized settler movement, or to "awe" the Indians, gave reason for revisiting the need for a regular army before and during the Constitutional Convention in 1787.[34]

Two competing ideas emerged during this period. Nationalists, including many former leaders of the Continental Army, held a Moderate Whig perspective. They desired a militia regulated by the federal government and subject to federalization when needed, a small regular army, and military institutions and establishments that would constitute the strategic and scientific elements of a national defense system—fortifications, depots, foundries, armories, and military schools. The anti-Nationalist position, espoused by numerous state representatives, held Radical Whig sentiments that saw exclusive reliance on the states' militias and supported no federal military infrastructure or institutions dedicated to military education.[35]

Neither idea prevailed in the 1780s. The Constitutional Convention eventually recognized the need for a permanent military establishment to protect the northwest frontier, and future contracts for seven hundred soldiers serving in that region came under federal authority. The convention also recognized the advantage of granting federal authority over key military depots such as West Point, and other "needful buildings." However, beyond a frontier force and limited infrastructure, the convention resisted the creation of a federal military establishment. Debate over authority to federalize the state militia remained unresolved because of state suspicion and the fear that high taxation would be needed to pay for defense.[36]

As a governing document, the constitution of 1789 enshrined a division of powers between the executive and legislative branches representative of compromise between the competing ideas. The president could not do what European sovereigns routinely did: start war regardless of public sentiment. The constitution granted the president authority as commander in chief of the army and the navy and of the militia of the several states "when called into the actual service of the United States." However, Article 1 affirmed that only Congress had authority to declare war and "to raise and support armies . . . [and] to provide for calling forth the militia." Congress also retained authority over the establishment and maintenance of military infrastructure and institutions—the means by which a nation stays prepared for war. The mechanisms for war preparation therefore remained at the prerogative of the people's representatives and not the executive branch. The constitution gave no provision for any academies or schools of military instruction. Supporters of the constitution called themselves Federalists and included most of the preconvention Nationalists.

The First and Second Systems

Conventions and constitution aside, once the Federalists gained power in 1790 they moved to create an army. The first secretary of war, John Knox, quickly tabled a proposal to federalize state militias in the Organization of the Militia Act of January 1790.[37] The expense associated with this proposal ensured its defeat in Congress.[38] However, Knox continued his efforts in the Uniformed Militia Act of 1792, which enshrined the traditional American ideal of universal service but also directed that all state militia be subject to federal service if called upon by the president. Knox demanded that all militia be trained using von Steuben's *Blue Book* regulations. While Congress passed the act, they included no mechanism to regulate reforms. There was no inspector general to ensure adherence to standards, nor were the states forced to apportion money to ensure sufficient training for the universal militia.

The disaster of St. Clair in 1791, the Whiskey Rebellion, and threat of war with England in 1793 forced change. Washington invoked executive powers under the 1792 Militia Act to federalize militias for Gen. Anthony Wayne's campaign on the Ohio frontier and for the president himself to deal with insurrection in Pennsylvania. To provide assistance to the coastal states, Washington turned to Congress, where they debated two options: investment in a large navy or construction of fortifications at select coastal sites.[39] Knox, heavily influenced by L'Enfant and Duportail, persuaded Congress to build forts, and in May 1794 it passed an act directing states to invest in coastal fortifications and authorizing the federal government to establish a corps of artillerists and engineers to assist in maritime defense. The act also created federal arsenals at Springfield, Massachusetts, and Harpers Ferry, Virginia, as well as a number of depots and armories.[40]

The fortifications were to protect major ports and harbors along the eastern seaboard from Portland, Maine, to Savannah, Georgia, while the government built key armories and depots inland to prevent their easy seizure by raid from the sea. The coastal fortifications, in what was subsequently called the First System, consisted not of masonry but of "parapets and batteries and redoubts . . . formed of earth."[41] In total, federal artillerymen mounted two hundred cannon on sixteen of these forts—half of them provided by the federal government, half by the states. The fort's garrisons housed 698

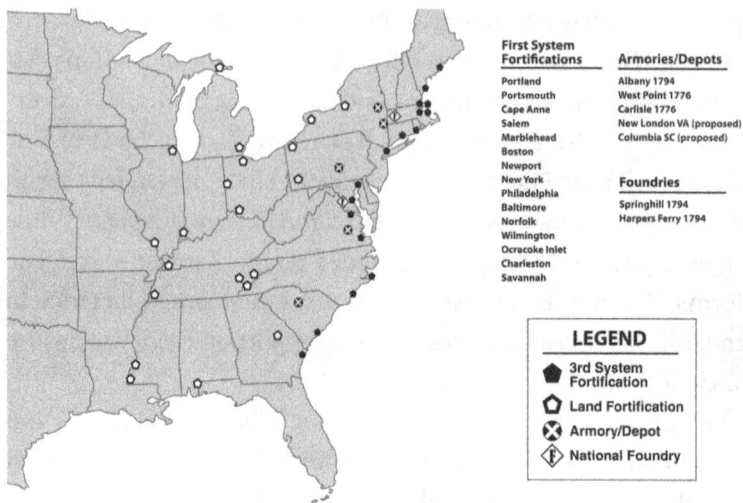

MAP 1. First System infrastructure, ca. 1800. Information compiled from "Fortifications," ASP MA vol. 1, no. 13 (February 28, 1794), 61. See also vol. 1, no. 27, (March 25, 1796), 172, for organization of the military establishment.

regular soldiers. The "Objects" of the regular military establishment were to occupy posts and "preserve peace" on the frontiers.

Work on these forts was only just under way when the Jay and Greenville Treaties brought the promise of peace and temporary respite for more congressional debate regarding the $2.7 million costs of the militia and coastal defense acts. At the same time, the War Department, without means to oversee the erection of seacoast defenses, passed responsibility for fortification construction to the individual states, who in turn hired European engineers to plan and oversee the work. Inevitably, political whim and neglect through transitions of government and fluctuation in appropriations diluted efforts to institute the full system. The results were therefore irregular and for the most part inadequate. Visiting Duke de la Rochefoucault-Liancourt summarized the accomplishments of the First System in 1799 thus: "They are either good for nothing or at least defective, so that money spent . . . may be said to be thrown away."[42]

The Quasi-War with France in 1798 renewed threats of European invasion and prompted the federal government to ask each state to repay any remaining debt incurred during the Revolution by committing the same amount to fortifications on their coasts.[43] The threat of war also allowed Federalists to justify temporary federalization of militia and their aggressive use against political opponents during Fries's Rebellion. By 1800 the Federalists had achieved almost everything that the Nationalists had proposed in 1783: western frontier forts garrisoned by 2,500 regulars, the power to federalize the militia, a federally controlled system of arsenals and military depots, state-supervised coastal fort construction, and a corps of artillery and engineers. What was still missing was an academic institution dedicated to the propagation of military science.

While Congress debated the expenditures for coastal defenses and the mobilization of state militias, military science was in limbo. This started to change with the 1794 act establishing a corps of artillerists and engineers and with a concurrent authorization for a number of "cadets" to undergo specialized instruction in the scientific branches of war. A Frenchman and veteran of the Revolution, Lt. Col. Stephen Rochefontaine, assumed command of the garrison at West Point in 1796 and began teaching Vauban's science of fortifications to selected young officers. Cadets and officers alike did not receive Rochefontaine well, and when a fire destroyed much of the school's infrastructure in April 1796, instruction tapered off and stopped after only a few months. Another four years passed before opportunity came again to establish a proper military school.

Alexander Hamilton became senior officer of the U.S. Army during the Quasi-War, and voiced dissatisfaction with progress of the First System and the poor state of officer training. In 1800 he proposed the establishment of federal institutions dedicated to tactical and scientific instruction. He envisioned a system of military education that included a common "fundamental school" at West Point and three specialist schools: one for infantry and cavalry officers, another for engineers and artillery, and a third for naval officers.[44] Hamilton persuaded Secretary of War William McHenry, who worked to convince President Adams and Congress of the need for at least a school for engineer-artillerists because "a degree of education and study was necessary in that corps, which was not required in any other."[45] Secretary McHenry argued that "no sound mind, after a fair view of the subject, can doubt the essentiality of military science in time of war" and that in times of peace, an academy was needed to instruct "the principles of war, the exercises it requires, and the sciences upon which they are founded." McHenry and Adams believed that military art consisted of different scientific branches that together allowed one to acquire a "perfect knowledge" of war, the possession of which "in its most improved and perfect state, is always of great moment to the security of a nation."[46] McHenry added that the United States should qualify officers "to place the country in a proper posture of defence, to infuse science into our army, and give our fortifications that degree of force, connexion, and perfection."[47]

President Adams suggested that because Americans would never agree to a larger standing army, then at least "as much perfection as possible be given to that which may at any time exist." He asked Congress to support the establishment of an academy where "military science in all its various branches ought to be cultivated" so that the nation would have ready a small number of very highly trained individuals whose purpose was to "diffuse and impart" their knowledge to citizen soldiers in an emergency.[48] At this same time Maj. Henry Burbeck, who had been stationed at West Point under Rochefontaine, assumed command of troops on the eastern seaboard and persuaded the secretary of war to allow him to send young officers to West Point to receive artillery instruction from experienced officers and civilian teachers. Burbeck's curriculum focused on mathematics, artillery ballistics, and Vauban's fortification system.[49]

Adams's desire to establish a military academy and Burbeck's West Point school might not have survived the elections of 1800. Adams's opponent, Thomas Jefferson, had long opposed the idea of military colleges, fearing that officers with Federalist associations would dominate them.[50] However, once in power and facing the possibility of renewed conflict with England, Jefferson approved the designation of Burbeck's school as the U.S. Military Academy. An act of Congress confirmed this on 16 March 1802. The incessant threat posed by England's navy during the Napoleonic Wars allowed the act to pass easily. Representatives from the Atlantic states were especially willing to support a military school that produced engineers and artillerists who might better construct fortifications along their coasts. In order to ensure that this academy did not become a school of Federalist persuasion, Jefferson directed that it be run by the U.S. Corps of Engineers, with the chief engineer (selected by Jefferson himself) acting as its superintendent.

It is easy to read too much into the formal establishment of a military academy in 1802. In reality, the event occurred without fanfare.[51] The announcement of Congress only changed the name of Burbeck's school. His handful of cadets and instructors carried on doing what they had begun the previous year. Day-to-day activity at West Point changed little, and in terms of its higher purpose, it was only a vocational school for specialist officers, a Woolwich on the Hudson River. The curriculum was rudimentary, including basic mathematics and geometry enough to understand Vauban's system of fortifications and defensive artillery applications. Strategy was not taught there, and Americans would not turn to West Point to find their next military commanders.[52]

Despite the limitations in the academy's curriculum, the beginning of the Jeffersonian era did nonetheless mark a turning point in the current of military thought in the United States, and it moved steadily away from British ways. The insistence on neutrality during the Napoleonic wars left the United States isolated, and as links with England diminished, so did English ideas regarding the preeminence of martial genius. Under Jefferson, Americans began to embrace military science instead. With its establishment, the West Point academy slowly became the formal center of American military thinking, which was becoming decidedly scientific.

A catalyst for the shift was Lt. Col. Jonathan Williams, nephew and personal secretary of Benjamin Franklin, a civilian scientist Jefferson commissioned in 1802 and appointed as the first superintendent of the military academy and chief of the new corps of engineers. One of Williams's first initiatives in November 1802 was to establish, at West Point, the U.S. Military Philosophical Society, enrolling all cadets and officers stationed there as permanent members. Williams used his secretarial position within the prestigious American Philosophical Society to recruit prominent citizens into his new institution, eventually attracting the active participation of hundreds of the nation's top politicians and soldiers, including three presidents, two future presidents, a vice president, three cabinet ministers, a chief justice, six governors, and thirty-three congressmen and senators.[53] The society thus became a network for lobbying for support to the army.[54]

The society, established "for the purposes of improving and disseminating military science and history," reflected a growing trend of gentlemen associations dedicated to mutual education and to the social advantages of affiliation."[55] In the main, the topics covered in societal papers dealt with fortifications, naval and weapons technology, and surveying science. The official seal of the society was symbolic of the emerging American view of war. During a society meeting at the War Department in Washington in 1808, Williams gave a description of the seal, recorded in the published minutes. It depicted the Greek goddess Minerva in armor "causing an olive branch to start from the ground by the touch of her spear: the goddess of science, in full armour, produces peace by the very lance which she is prepared for battle."[56] The accompanying motto read *Scientia in Bello Pax* (through science in war, peace). Williams's society and his aim for the new academy demonstrated his desire to create an army "officered by men well educated in every branch of military science."[57]

In the society's papers, as in so many of the writings of the period, one can discern a predilection for rationalism and a continued belief in the utility of science and progress. These were common Enlightenment ideas basic to humanitarianism, deism, rationalism, and utilitarianism, all promoting a faith that progress was the natural path for humankind and that science was the guide to progress. Science, to these thinkers, incorporated not only the physical sciences but also what is now distinguishable as technology. American military thinking emerged in the young republic solidly committed to the pursuit started by les Lumières—the discovery of scientific components of war, with complete faith in the power of reason and with an unprecedented belief in the utility of mathematics as key to all scientific endeavors, which could grant a perfect knowledge of war.[58]

Despite the early success of his society, Williams's goals for the military academy were constrained by a lack of instructional expertise and continuity of cadet attendance, want of proper texts, and a narrow mandate to train only a small number of officers. Williams envisioned the academy not as "a little mathematical School, but a great national establishment," the equivalent of the French École Polytechnique—an institution producing a scientifically trained bureaucrat and civil and military engineer—"an *Officeur du Genie.*" Williams thought this type superior to the English concept of an engineer: "I cannot find any full English Idea to what the French give to that profession." He desired most of all to create an elite: "We must always have it in view that our Officers are to be men of science, and as such will by their acquirements be entitled to notice of learned societies."[59]

Despite his grand vision, for the first decade of the military academy's existence it remained only "a little mathematical school." The 1810 *Regulations*, for instance, outlined the entire curriculum in a single paragraph, and a cadet could master this in less than a year.[60] No standard for the length of studies or course content existed. Until significant reforms during the period 1817 to 1820, the academy was elementary in terms of its scientific instruction, and it focused exclusively on the provision of knowledge to a small number of officers. An English legacy still endured that saw military science as essential only for the military specialist, the engineer and the artilleryman, and of little value to the infantry or cavalry officer. Therefore, despite Williams's musing over creating a scientific officer corps, the academy at West Point was similar to many of the private military schools that existed throughout Europe at the time.

There was, however, a growing demand for change. The *Chesapeake* incident of 1806, followed by the threat of British blockade of Hampton Roads, Virginia, fueled the initiation of the Second System of defense. In 1806 the Senate commissioned a progress report on the defenses of the First System which determined that of the original twenty-one fortification sites, only sixteen had been constructed to any degree, and most of these were primitive and ill equipped to withstand assault. Jefferson and the Republican Congress then appropriated $1,927,100 for the construction of permanent masonry defensive structures at forty-four critical harbor and river mouth locations and commissioned plans to build dozens of small gunboats to provide mobile gun platforms to cover gaps between permanent coastal defensive works.[61]

As construction commenced in earnest on this Second System, want of proper engineering and artillerist skill and general military science became obvious. This want, combined with fear of war with England, pushed Congress to authorize changes to the academy in 1812. On 29 April they passed an act that converted its mandate from one focused on the education of specialist officers to the provision of military education for officers from all branches of the regular army. In a significant departure from European practice, the congressional act directed that "cadets heretofore appointed in the service of the United States, whether of artillery, cavalry, riflemen, or infantry, . . . be attached at the discretion of the President of the United States, as students to the military academy."[62]

This legislation broadened the role of West Point widely, making it the only common-to-all-arms national military academy in the world. The president, through the office of the secretary of war, appointed cadets, ending the practice of regimental or corps sponsorship, with its lack of system for appointments and promotions. Cadets would receive an affiliation with a particular branch of the

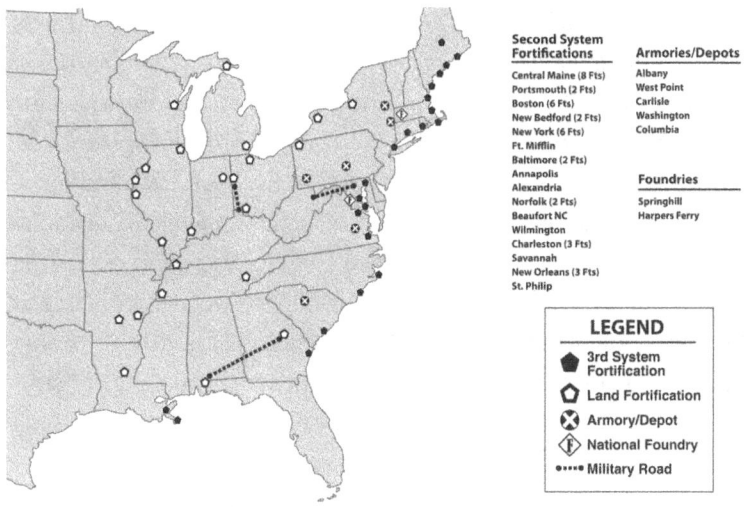

MAP 2. Second System infrastructure, ca. 1812.
Information compiled from ASP MA, vol. 1, no. 59, 191.

army only upon graduation. The management of the academy continued under the corps of engineers. Studies ceased during most of the War of 1812, and when they recommenced, the idea of a purely scientific curriculum was sustained. All cadets, regardless of branch, received a lengthened and increasingly more scientific education, one designed to allow all regular officers to oversee defensive construction and military ordnance and logistical preparations for war.

Military Science and the Militia Officer

As the U.S. Military Academy became committed to teaching fortification and artillery applications to regular army officers, a few militia officers began to write treatises reflecting their own interest in military science. Since the Revolution, these officers had shared personal libraries and sought to acquire military dictionaries and encyclopedias from Europe for reprinting in the United States.[63] Then, in 1809, Lt. Col. William Duane produced the first American rendering of military science, *The American Military Library—or Compendium of the Modern Tactics, Embracing the Discipline, Manoeu-*

vres, and Duties of Every Species of Troops, Infantry, Rifle Corps, Cavalry, Artillery of Position, and Horse Artillery; A Treatise on Defensive Works in the Field, the Exercise in Sea Coast Batteries, and Regular Fortifications, Adapted to the Use of the Militia of the United States. This was, as the title suggests, a thorough examination of the history of war, tactics, and "greater tactics." Duane's work was a remarkable synthesis of two hundred contributing sources, written in two volumes and presenting "the most approved system" of war, which was "the French system." He opened with a history of warfare and ended with an analysis of the modern campaigns of Napoleon as described by the then little known Antoine Jomini. As such, Duane's analysis was very contemporary.[64]

Duane—himself a friend of Jefferson and a successful publisher—wrote for American militia officers, men of politics, and educated men interested in the emerging science of war. His work was comprehensive, covering how basic geometry applied to field fortifications and the application of artillery fire, ordnance functions, the organization, characteristic, and tactics of all arms, staff organization and procedures, and the training and supplying of armies. However, his most current subject was "strategy," which he called the "science of commanders."

The idea of a science of commanders had emerged in 1754 in Le Comte Turpin de Crisse's *Essai sur l'art de la guerre*, in a chapter entitled "A Principle on which the Plan of Campaign May be Established." De Crisse used Vauban's siege parallels to explain how an army should advance from point to point deliberately on offensive campaigns. This influenced Henry Lloyd, who posited in 1780 that armies follow "lines of operation" between key points.[65] European writers then began to use geometric models to explain campaign maneuvers. By 1800 there were several systems of campaign design published in French and German works. Duane gave particular praise to what he referred to as the first modern system, "the system of the base of operations."[66] He told his readers that real change was occurring in war in Europe where the tactics of armies was established upon the new principles of science. The mathematics which had with the aid of the magnet, so much enlarged the scope of human curiosity, knowledge, and power, was soon employed in the art of war; in the management of artillery, the erection of military works, and in fixing certain laws for the management and direction of armies.[67]

To Duane the most important of these laws concerned selecting a base of operations and a sound system of logistics: "Thus the necessity of subsistence, and of ammunition, gave rise to the principle of the base . . . the establishment of certain points where magazines are established; whence armies move, to combine their attacks against an enemy."[68] Duane introduced into the American lexicon terms and concepts that were to have great import on nineteenth-century warfare. They were not the writings of Jomini but of the German theorist Heinrich Dietrich von Bülow, who in 1799 defined a geometrical concept of military strategy in his *Geist des neuern Kriegssystem*, including concepts that were to have significant impact on the American understanding of strategy.

Bülow proposed a formal system of warfare that reduced its chaos and promoted understanding. The key element of his system was the "science of movements," divided into two parts that together constituted the art of war.[69] The first he called "tactics," which he described as movement within reach of each side's artillery. The second part he called "strategy" and incorporated all movement of armies out of "the visual circle of each other . . . out of cannon-reach."[70] Such movements were made within a campaign plan that required generals to select a base and lines of operations that together formed—optimally—a triangular area of operation within which an army could maneuver and fight.[71] Bulow's fundamental principle for strategic offensive campaigns was an evolution of Turpin's and Lloyd's theories and was divided into three essential parts, a *base of operations* (the base of the triangle), an *objective point* in enemy territory, and *lines of operation* from the base to the objective (see fig. 1).[72]

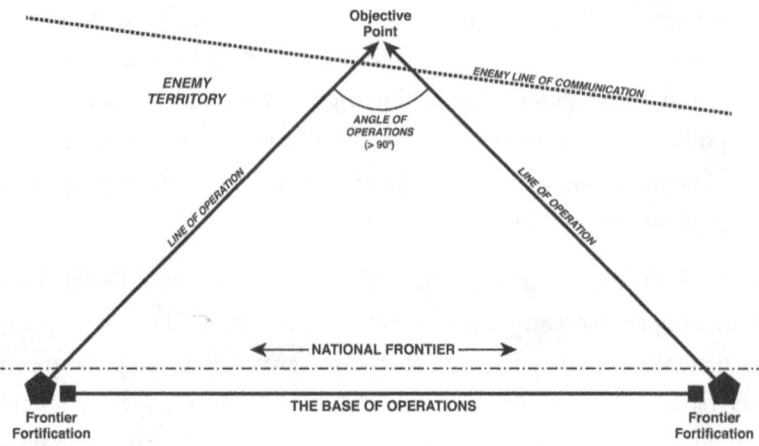

FIGURE 1. Heinrich von Bulow's Optimum Base of Operations. Based on *The Spirit of Modern System of War* by Heinrich von Bulow (London: T. Edgerton, 1808) 23–24, 183; and *The American Military Library* vol. 1 by William Duane (Philadelphia: William Duane, 1809), 70–72, 78–82.

The establishment of the base of operations would be as Vauban had originally directed—along a frontier—secured by fortifications that served as depots from which an army might march into the enemy's territory.[73] The advance from the base was projected upon a line of operation toward an objective point that cut the supply line (the line of communication) of an enemy's army. Optimally a line of operations could be projected from either end of the base of operations toward the same objective point, forming a triangle. A commander could then conduct marches into enemy territory in such a way that the army was never more than three days from a supply point, either along the base of operations or along a line of operations.[74] The establishment of supply magazines along this line could ensure "velocity" of maneuver. To achieve success, one had to ensure that geometric lines drawn from each end of the base of operation to the object form an angle of at least 90° with the enemy's main army outside of the triangle thus formed.[75] The key concept here was that the attacking army should have space enough within this triangle to maneuver to the flank of an enemy army and cut its line of supply; if forced to withdraw, it could do so to a point on the base least threatened and could withdraw without threat of being itself flanked by the enemy army.

A base of operations that was too short, combined with a line of operations that was too long, would create an altitude in the triangle such that the angle of operations was very small, meaning simply that the army had limited space to maneuver. An angle of operation less than 60° or a base of operation that constituted but a single point along a frontier was considered dangerous, as an enemy could outflank an army and sever it from its base.

The influence of von Bülow in Europe was significant, and subsequent European works on military science, including Jomini's, used parts of his system. In fact, Bülow's work provoked a competition between Jomini and Archduke Charles of Austria regarding what was valid and what was ridiculous about the geometry of military movements. Bülow had compromised his own work with pedantic descriptions of convex and concave bases and convergent and divergent lines of operation. Jomini carried on the competition and the compromise, providing complex combinations that today appear nonsensical because they ignore the idea that all operations must conform to the dictates of topography. Duane did not replicate these complex geometric combinations, and his explanation of Bülow's strategy was more subdued, setting the tone for subsequent American presentations of the system at West Point, where the essence of the theory of the base of operations endured throughout the antebellum period within a course entitled "the science of war."

Another militia officer, Brigade Maj. Epaphras Hoyt of Massachusetts, wrote *Practical Instructions for Military Officers* (1811). This was a re-publication of Roger Stevenson's *Military Instructions for Officers Detached in the Field* (originally published in 1775) with three distinct editorial changes. The first was an extensive improvement on the first chapter containing elementary geometry. The second was a new chapter on fortifications. The third was Hoyt's addition of strategic science. He was the first to offer the American reader distinctions between the three "principal branches" of war: tactics, engineering, and strategy. Tactics he defined as "the rudiments of discipline with the arrangements, movements, and manoeuvres of troops in the field." Engineering was "the mathematical, philosophical, and mechanical parts, including fortifications . . . and the whole science of artillery." Strategy, according to Hoyt, "embraces the science of command, or the duties of generals upon the extended scales." Hoyt clearly based his strategy on Bülow's science of movement. He reiterated Duane's lament that militia and regulars alike neglected all three branches of the military art in the United States for want of proper study.[76]

While it is hard to gauge, the impact of Duane and Hoyt's works does not appear to have been great. Published not long before the commencement of the War of 1812, they were not so widely disseminated as to influence militia performance during the war. However, both works did influence secretary of war John Armstrong, who in 1812 decided to produce his own handbook on military science. This short, accessible treatise is notable for its reduction of all military science to two overriding principles: concentration of military force and celerity (speed and momentum) of movement. This handbook was decidedly Napoleonic.[77] Little evidence indicates that Armstrong was widely read, either. His work, as well as Duane's and Hoyt's, probably serve historians more than they served militia officers of the day. They are valuable in that they provide us with a snapshot of emerging military thought among some militia officers in the Jeffersonian era, giving clear proof of some interest and inclination toward systematic thinking about war and appreciation for military science. They also show that officers in the United States were capable of Americanizing European ideas.

The colonization of North America did not include the transference of military science. Before the Revolution, colonial military knowledge was rudimentary. Military preparedness involved an Elizabethan tradition of universal militia service, augmented by volunteer companies raised by colonial elites. While colonial military thought was decidedly British in 1775, French military science grew short roots in the young republic during the Revolution, and Nationalists and Federalists kept it alive thereafter. It gained currency when Congress passed legislation to fortify coastal frontiers and establish a corps of engineers and artillery in 1794. Establishment of the military academy in 1802 provided a place for instruction in military science, but it took the War of 1812 to complete this institutionalization.

2

Army Reforms, 1815–1820

The majority of citizens of the United States in the Jeffersonian era belonged to the Atlantic world. Wealth and well-being relied on maritime security. However, the War of 1812 exposed the vulnerability of the Atlantic and Gulf coasts to invasion. First and Second system defensive works proved adequate for local defense but did nothing to prevent the Royal Navy from wreaking havoc on unprotected points in Maine, the Chesapeake, Georgia, and the Gulf. The independent-mindedness of each state and a highly decentralized war management system prevented a coordinated effort to counter this threat. Tired of such inefficiencies, President James Madison and Secretary of War James Monroe pushed for organizational and doctrinal reform in 1815. Congress established permanent staff bureaus in Washington, and a system of coastal fortifications and federally controlled mobilization evolved into a coherent national defense policy in 1821. The regulation of military affairs arising from these reforms paralleled a complete overhaul of the military academy under Superintendent Sylvanus Thayer. New regulations, Napoleonic doctrine, and adoption of the French pedagogic system provided mechanisms for the professionalization of the officer corps that commenced in this period.

The War of 1812

After twenty years of tension between the United States and Great Britain over freedom of navigation on the seas, war erupted in June 1812 and American coastal frontiers immediately became vulnerable to amphibious attack. The U.S. Navy remained too small to confront British squadrons in open waters. Ports in Nova Scotia, Bermuda, and the Bahamas provided bases from which the Royal Navy could threaten any point along the east coast of the United States. In the spring of 1813, the naval forces of British admirals George Cockburn and John Warren blockaded Chesapeake Bay and harassed coastal towns in Maryland, Delaware, and Virginia. Where American militias erected even small earthworks, Cockburn shelled local villages. Where militia garrisons encamped, Cockburn landed forces to exact punishment upon the nearby inhabitants. Cockburn's exploits were soon in every American newspaper and on the minds of people living on the bay, as the British plundered Havre-de-Grace, Frenchtown, Georgetown, and Fredericktown with as much brutality as efficiency.[1] The Second System defenses did nothing to prevent British domination of major maritime channels.

In late June 1813 Cockburn and Warren concentrated twenty vessels, including fourteen ships-of-the-line, with thousands of troops aboard at the waterway of Hampton Roads, threatening the port of Norfolk, Virginia. Well-placed earthen fortifications and seven well-sited cannon thwarted a British attack upon "a forward defensive work" on Craney Island on 22 June, killing and wounding more than 200 of the 1,500 assault force.[2] The British exacted vengeance two days later when they captured the unprotected hamlet of Hampton, across the James River from Norfolk. National papers quickly picked up reports of widespread rape and property destruction and told the shocking details of how the women were "literally hunted down" by gangs of looting sailors and soldiers.[3] The British authorities did not deny the claims, and shock waves resonated through eastern Virginia as Warren's naval forces lay at Hampton Roads until August 1813.[4]

The threat of naval raids reemerged in the summer of 1814 as Royal Navy squadrons again descended upon the Chesapeake. People abandoned Hampton, and some inhabitants of Norfolk fled. However, Norfolk was prepared for the renewed threat. That spring, Captain of Engineers Sylvanus Thayer, graduate of West Point, arrived to coordinate the defenses of the town, improving upon the works made the year before.[5] Those who had remained in Norfolk breathed easier when British forces passed them by and moved up the Chesapeake to the Potomac and Patuxent. There the British routed American forces at Bladensburg, sacked Washington and Alexandria, and burned national public properties. They then maneuvered by both sea and land toward Baltimore. As with Norfolk, however, Baltimore had not ignored its defenses. The city had mobilized its citizenry, organizing an effective militia muster and erecting extensive earthworks. These, together with the substantial Second System fortifications of Fort McHenry, stemmed the British tide. Yet the victory did not compensate for the awareness of how vulnerable these areas were to ships of foreign powers.

The pillaging on the Chesapeake in 1813 and 1814 was a formative American experience. It induced widespread fear. Most of the population of the United States still lived upon the littorals, and maritime navigation and trade were of first importance. This was especially so for those who lived on the Chesapeake, where continued British blockade threatened the livelihood of businessmen and where violence wrought during the war left an indelible impression.[6] People wanted protection of their coastal towns and their maritime enterprises, and they came to realize that the threat of invasion was much greater on the maritime frontiers than on the boundary with British North American colonies. The English occupation of portions of the coast of Maine and Georgia and landings at New Orleans and Mobile left no doubt about this.

The possibility of further attacks concerned the politicians who traveled through Washington in September 1814, past the burnt-out shells of the government buildings, to meet under presidential summons in the Patent Office, the only federal building left intact. In the congressional debate that followed, James Monroe, now secretary of war, called for improvements in the ordnance bureau, increases to the corps of engineers, and an increase in the size of the regular army. He asked for 40,000 more regulars for garrisoning coastal defenses to "relieve the coast from the desolation which is intended for it." He proposed the creation of another force of 100,000 regulars to capture the Canadas, in order to "keep in our hands a safe pledge for an honorable peace." The most contentious issue, however, was his call for a universal conscription under federal control (but at state expense). Monroe wanted sufficient forces to deny the British their avowed purpose, "to lay waste and destroy our cities and villages, and to desolate our coast . . . to press the war along the whole extent of our seaboard, in the hope of exhausting equally the spirits of the people and the national resources."[7] Neither the president nor his secretary of war believed the states were strong enough separately to deal with this threat.

Congress rejected these proposals, supporting instead an increase in short-term militia for local service to defend threatened coastal regions. State representatives saw in Monroe's suggestions the desire to create a large standing army, which indeed it was. However, Monroe was not attempting to secure that which Hamilton once coveted, an army for social control. Rather he wished to correct a thing that no one could dispute: the extreme inefficiency of state-controlled mobilization.[8] At no time was the British force on the northern frontier greater than 14,000 regulars, militia and Indian fighters combined. At sea, the British maritime contingents never mustered more than 14,000 sailors, soldiers, and marines. At the same time, the United States employed 60,000 regulars and 471,000 militiamen during the war (not more than 235,000 at any one time). And the expense of a military establishment based on a short service militia system was staggering.[9] Yet, in the autumn of 1814, all of this seemed woefully inadequate.

Suspicion of Monroe's design was balanced by continued threats posed by British forces that had already sacked Havre-de-Grace, Frenchtown, Georgetown, Fredericktown, Hampton, Bladensburg, Washington, parts of Maine, and Cumberland and St. Mary's Island on the Georgia coast. Just as the states were questioning the government's ability to provide for a common defense without widespread federalization, news arrived of the U.S. Navy's victory at Plattsburgh. This and the subsequent success of Andrew Jackson at New Orleans prevented further invasions. Victories and news of war's end lent a triumphant luster to the spring of 1815, and Americans unaffected by the pillaging of the Chesapeake forgot the war and basked in rays of newfound nationalism. Apparent success, however, did not blind the new secretary of war.

Military Reform and the Third System

Having failed to secure conscription, Secretary Monroe was still determined to reform the military establishment. The war had exposed want of a proper general staff system. In 1812 the secretary of war himself had attempted to function as a chief of staff to coordinate state efforts. However, dealing with eighteen state militia systems proved overwhelming, and he distributed the tasks to newly formed staff bureaus. Initially these included adjutant and inspector generals, a paymaster general, and a chief of ordnance, but expanded to include a quartermaster general and a commissary general. The secretary of war also brought the chief of engineers, the only congressionally approved army staff position, to Washington from West Point to take charge of a newly formed Engineer Bureau. Eventually he also created a bureau of Topographical Engineers and Judge Advocate and Surgeon General's offices. These staff bureaus supported the office of the secretary of war directly, coordinating activities on his authority with the states and with generals commanding the geographic departments (sometimes called divisions). Congress sanctioned these staff bureaus in 1816, formalizing approval in the Staff Act of 14 April 1818, thereafter guaranteeing congressional funding.[10] The staff bureaus created standing boards whose reports passed through the office of the secretary to the congressional Committee on Military Affairs.

The new staff arrangement (kept in existence until well into the twentieth century) set in motion an era of greater regulation of all American military affairs and ended an era of reliance on personality and idiosyncrasy. A senior regular army commander, Brig. Gen. Winfield Scott, reinforced this effort by publishing his comprehensive (355-page) *General Regulations for the Army* to replace von Steuben's *Blue Book*. These regulations set down how the army would follow the direction of the staff bureaus using standard forms and procedures for reports and returns for officers employed as assistant quartermasters, assistant adjutants general, paymasters, or assistant inspectors general in the geographic departments and in subordinate units and posts. Brig. Gen. Thomas Jesup (who would remain army quartermaster general until 1860) ensured that his bureau scrutinized every report for proper accountability of dollars spent, a practice that continued throughout the antebellum era with the result of giving the army a reputation for honesty and efficiency in management.[11] Neglected by all but a few military historians, these staff bureaus and regulatory reforms constituted in fact an American general staff system. Military science eventually became their doctrine as the staff bureaus grew to depend upon the mathematical and scientific expertise of West Point graduates to fulfill ordnance, survey, construction, quartermaster, and adjutant general functions.

Historians have criticized Monroe for not establishing this military bureaucracy under a single general in chief in Washington, who might command both the geographic departments and the staff bureaus. Indeed this would have enhanced military unity of command and created a Prussian-style general staff. Such thoughts, however, were unpalatable in the Jeffersonian era. American insistence upon division of power and congressional authority applied equally to military as to political affairs. The state representatives in Congress remained suspicious of any supreme general officer and preferred to divide military power. Monroe's decisions and actions were thus constrained by traditional American principles of governance and by a legacy censuring large standing armies. The 64,000 regular soldiers in service in February 1815 were reduced by congressional act in March to just 10,000, whose purpose was to garrison frontier forts and, according to South Carolina congressman John C. Calhoun, to "keep alive military science."[12]

Forced to accept the small size of the regular army and the lack of authority to regulate and federalize state militias in times of peace, Madison and Monroe turned to other means to improve the military establishment. Creating general staff bureaus under control of the secretary of war and organizing the regular army into geographic department and district commands were important reforms. However, these organizational measures alone did not constitute a concept for the common defense or a governing strategy. What was needed was a fiscally feasible and politically acceptable national defense policy that could be managed by federal authority without threatening states' rights and which at the same time addressed the most obvious threats to American livelihood: the interruption of trade and invasion.

The answer came incrementally in dialogue between the president, the secretary of war, and the office of the chief engineer. All agreed on the need for an extensive system of permanent coastal fortifications, under federal control, as originally envisioned by General Duportail. Immediately following the cessation of hostilities, Chief Engineer Brig. Gen. Joseph G. Swift received approval from Secretary Monroe to send a team of engineers to inspect fortifications along all coasts to select "judicious sites for new works."[13]

The secretary also agreed to send a team to France—"the seat of war" in Europe—to acquire knowledge of the most modern fortifications science. There were few alternatives to France. In the aftermath of the "second War of Independence," the United States found itself further estranged from Britain and with few genuine friends. France was likewise alone and very willing to provide guidance to an old ally. This mission was to acquire expertise in fortifications design, military and civil engineering, army organization, and military education. In the wake of the Napoleonic wars, Americans (and most Europeans) held French military practices in all these areas in the highest regard. The United States was eager to reestablish solid and mutually beneficial relations with her partner of the Revolution and to acquire military doctrine and plans that could be of use in solving America's considerable national defense problem.

Monroe and Swift sent Lt. Col. of Engineers William McRee, West Point class of 1805, and young Capt. Sylvanus Thayer to France. Thayer had remained in Norfolk until war's end. He received credit and brevet promotion for his good service on the northern campaign in 1812 and for his "Scientific Defenceable Works at Norfolk." He left for France in the spring of 1815 with the explicit direction to become "conversant with all the latest European ideas on scientific schooling, on the theory and practice of fortifications," and to gain knowledge of "the military schools and work-shops, and arsenals, the canals and harbours, the fortifications, especially those for maritime defence."[14] McRee and Thayer were also given funds to buy books, maps, and instruments.[15] Thayer spent almost two years in Europe before returning to superintend the military academy in 1817 and to begin there a series of reforms to incorporate military science into the academy's curriculum.

While Thayer's mission to France was under way, President Madison, keen to ride the tide of nationalism, set the conditions for new coastal defenses by reconstituting the national bank.[16] He was forced to temper his desire in order to placate old Republicans who feared too much federal government involvement and westerners who did not share his Atlantic world concern. Still, although money was scarcer than previously, the government appropriated $838,000 for defensive improvements in 1816.[17]

In April of that year, Congress granted approval for the employment of Napoleon's own chief engineer, Simon Bernard, as a regular army brigadier general in charge of all of the nation's fortifications.[18]

They also granted permission to employ French instructors at the military academy. At the same time James Monroe, who had handed the duties of secretary of war to William H. Crawford (former American ambassador to France), commenced a trip to visit all coastal fortifications throughout the United States, in what was ostensibly a pre-presidential election campaign tour. However, he achieved the military intent and became convinced of the need for integrated fortified frontiers. In November 1816 he helped to establish a Fortifications Board in Washington under the leadership of Simon Bernard.

This board, consisting of Bernard and McRee (later joined by army engineer Joseph Totten), began surveying America's coasts to determine fortification requirements. Bernard communicated his vision for a system of fortifications immediately and clearly. He wanted it to be part of a broader defense strategy. The United States required what Vauban had given France in the seventeenth century: fortified frontiers connected to a system of mobilization, supply, and transportation that would allow state-sponsored militias to respond adequately to any maritime threat. Bernard wished to establish an American *ceinture de fer*. This marked the beginning of what was later called the Third System of defense, which was to continue for the next fifty years. It was also the beginning of an enduring need for military science.

Newly elected President Monroe publicly announced this new system in his inaugural address on March 4, 1817:

> Many of our citizens are engaged in commerce.... These interests are exposed to invasion.... To secure us against these dangers our coast and inland frontiers should be fortified, our Army and Navy ... kept in perfect order, and our militia be placed on the best practicable footing. To put our extensive coast in such a state of defense as to secure our cities and interior from invasion will be attended with expense, but the work when finished will be permanent, and it is fair to presume that a single campaign of invasion ... would expose us to greater expense.[19]

Monroe then expressed his opinion on the role of the army and militia within the system. The army was to "garrison and preserve our fortifications and to meet the first invasion of a foreign foe, and . . . to preserve the science as well as all the necessary implements of war." States were to organize and train their militia to defeat subsequent invasions by foreign armies.

The planning and construction of fortifications along the coasts affirmed that those in power still identified the United States as part of the Atlantic world. They gave some consideration to inland frontiers, but these were seen as corollary to Third System requirements. Simon Bernard disclosed this sentiment in a report he prepared for the new secretary of war, John Calhoun, in 1818:

> Three of our frontiers require special attention: the eastern, or Atlantic frontier; the northern, or Canadian frontier; and the southern, or the frontier of the Gulf of Mexico. On the west and northwest we are secure, except against Indian hostilities; and the only military preparations required in that quarter are such as are necessary to keep the Indian tribes in awe. . . . All our great military efforts . . . must for the present be directed towards our eastern, northern, or southern frontiers, and the roads and canals which will enable the Government to concentrate its means for defence, promptly and cheaply, on the vulnerable points of either of those frontiers. . . . The Atlantic . . . is the weakest and most exposed.[20]

Prolonged discussions between Secretary Calhoun and the engineer and ordnance bureau chiefs produced a consensus about Bernard's strategic concept. Captured in a detailed Fortifications Board report dated February 7, 1821, this concept anchored a comprehensive defense policy and set the course of military planning and preparations until disrupted by war in 1861. The report pointed out that the United States could not afford a seagoing navy sufficient to deter foreign invasion, and therefore had little choice but to prepare a small inshore navy and land defensive works for coastal protection:

A defensive system for the frontiers of the United States is therefore yet to be created; its bases are: first, a navy; second, fortifications; third, interior communications by land and water; and fourth, a regular army and well organized militia: these means must all be combined, so as to form a complete system.[21]

The coastal frontier was to have an integrated system of permanent fortifications that protected cities and harbors and key waterways: New Orleans, Norfolk, Baltimore, Philadelphia, New York, and Narraganset Bay (see map 3). Regular artillery troops would man the defensive works in peace, commencing a practice of maintaining proportionally high numbers of artillery soldiers in the regular army. In times of crisis, these fortifications would limit the places where foreign troops might land, slow down invasion, and buy enough time for substantial militia forces to muster and—using improved roads and waterways—move quickly to the threatened point.[22] The report put fortification construction costs at $11,147,695 and garrison requirements at 2,720 regular troops in peacetime and an additional 21,000 militia in times of crisis. Another 36,280 militia would be required in war to move to and counter any landed invasion force.[23] The report found these costs cheaper than might be incurred if dealing with invasion without the benefit of fortifications. It was estimated that expenses for mobilizing sufficient military personnel to defeat an invading army of 20,000 soldiers would equal $5,658,000 for a six-month campaign. Damage done by invaders would triple these costs. Calhoun endorsed this systematic plan and moved it through Congress.[24]

During the next four decades Congress dedicated considerable legislation and a sizeable portion of the monies appropriated for military matters to implementing this strategy of fortified frontiers, controlled by the Fortifications Board.[25] The board reported directly to the secretary of war and not to any military commanders, and this allowed the corps of engineers to achieve hegemony over military policy.[26] Senior engineers contributed more to debates

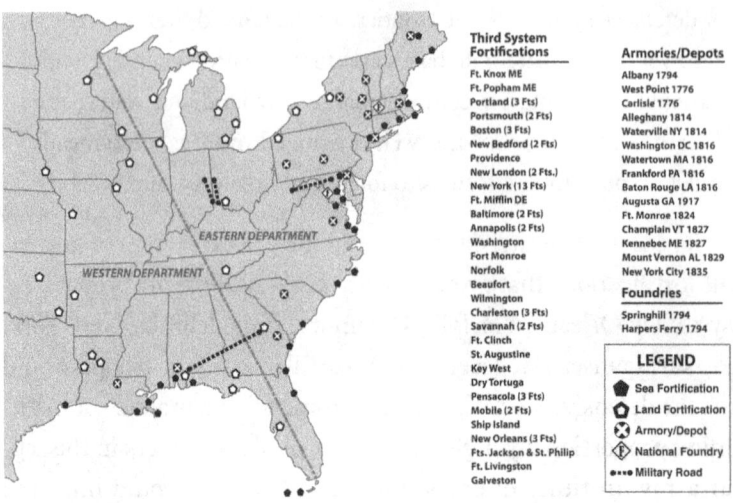

Map 3. Third System infrastructure, ca. 1821.

in Washington during these formative years (1817 to 1826) than did Brigadier Generals Winfield Scott and Edmund Gaines, the eastern and western geographic department commanders.[27]

Beyond fortifications themselves, another essential component of the Third System was the regular army, and its size and role mattered. Secretary Calhoun had sought to retain the 10,000 soldiers agreed to in 1815. But in 1820 a watchful Congress, suspecting that a larger military establishment (regular and regulated militia) would increase taxation, sought justifications for all regular army personnel and set out to reduce the army to only the garrison numbers required by the defensive plans being drafted by the Fortifications Board.[28] This concerned military leaders, particularly with regard to the officer corps. Secretary Calhoun directly addressed the issue in a report entitled "Reduction of the Army" in which he captured "the general principles on which it is conceived our military peace establishment ought to be organized."[29]

Calhoun outlined two purposes of the regular army. The first involved garrisoning "forts along our Atlantic frontier in order to . . .

cause the sovereignty of the United States to be respected" and the occupation of posts "on our inland frontier to keep in check our savage neighbors." The second purpose was to keep the army in a state well prepared so that "the dangerous transition from peace to war may be made without confusion." Therefore, in any reduction of the army, Calhoun requested maintenance of a proportionately larger regular officer corps that could form the leadership cadre needed for an "expansible army" created in emergencies through direct recruitment or federalization of state volunteers.[30] Calhoun reinforced his argument by stating the general purposes of militia, volunteer, and regular forces. He claimed that the militia was still "the great national force," but qualified this by saying that while "they may be relied on to garrison forts, and to act in the field as light troops . . . to suppose our militia capable of meeting in the open field the regular troops of Europe, would be to resist the most obvious truth." Calhoun wanted a small corps of regular troops and a disproportionately larger corps of regular officers trained in "the present improved state of the military science" who would lead large militia formations in times of war. This cadre of regulars could man defensive works and drill regular and militia soldiers, but their real *raison d'être* was to conduct "indispensable preparation" for the inevitability of war. In this he saw the military academy at West Point as an "invaluable part of our establishment."[31]

While Congress approved the Fortifications Board's report of February 1821, Calhoun's proposal of an "expansible army" did not receive widespread support. But he enacted it anyway, and this, together with Third System requirements, directly affected the West Point academy.[32]

Reforming the Military Academy

West Point was in want of reform after the War of 1812 to set it up to fulfill the policies of the secretary of war. Only one officer graduated in 1813, but more than 150 cadets turned up for instruction the following year and more the year after. At this time desire outpaced capacity, and the staff and faculty of the academy fell into upheaval trying to make do. During the war the management of West Point had been left to Alden Partridge, a later much-maligned captain of engineers. After graduating in 1806, Partridge had never left the academy. As acting superintendent from January 1815 until July 1817, Partridge used what faculties a man inexperienced in military applications could muster to keep the academy growing and improving. He was served well by retired lieutenant colonel of engineers Jared Mansfield, professor of natural and experimental philosophy, and esteemed surveyor Andrew Ellicott, professor of mathematics, who both attempted to bring a semblance of organization to the academy's curriculum and to invoke uniform standards. But it was not until intervention by Secretary of War William Crawford with the issuance of a new set of regulations in 1816 that things really began to change.

The following year marked what one historian of military education called the commencement of a "scientific culture" at West Point.[33] The new regulations stipulated four key things: that the academy follow a pattern of lessons from September to the end of June, with examinations in December (later moved to January) and June; that a Board of Visitors of five "gentlemen versed in military and other sciences" convene to observe the June examinations; that cadets be admitted each September only after having passed an entrance exam; and that cadets follow a four-year course of studies covering all the branches of science necessary for mastery of the art of war and fulfillment of the nation's defense requirements. It took three more years to hammer out the exact specifications of this new curriculum. Instrumental in doing this was Sylvanus Thayer, Mansfield, Ellicott, and French professors Christian Zoeller, Claudius Bérard, Pierre Thomas, and Claudius Crozet. Of these, Crozet was most influential. He was a graduate of the École Polytechnique and had served as an *Officeur du Genie* in Napoleon's Grand Armée through many of its most famous campaigns, including Waterloo.

Crozet was determined to reform the curriculum and to make its content modern and French.[34]

Thayer himself agreed. He had arrived at West Point in August 1817 full of ideas after two years in France, where he had observed professional instruction at various institutions. He brought with him hundreds of books and plans, models, and ideas for course content and pedagogy. Thayer's first challenges involved personality differences with outgoing superintendent Partridge and disciplinary problems related to cadet conduct. Thayer made the cadets sit through classification exams to determine their individual standing before moving to establish a new four-year curriculum and placing each cadet into a year group. He also implemented the new set of regulations that would make cadet life highly structured. A number of cadets could not handle these disciplinary reforms, many of them having come from households where personal honor and a strong sense of individualism stood above all things in importance. These cadets were outraged when Thayer demanded that they submit to uniform standards of conduct.[35]

The reforms included a new governance structure. The chief engineer had initially run West Point. After 1812, when the chief engineer moved to Washington, he appointed a superintendent to manage the academy. In practice, the position of superintendent carried with it full power with regard to academy affairs, academic and disciplinary. The superintendent reported solely to the chief engineer, who himself reported directly to the secretary of war. There had never been interference in academy affairs by military field commanders. This aspect of the system remained in place.

Within the academy, however, reforms in 1817 put responsibility for academic and disciplinary decision making with the newly formed Academic Board. This board was composed of the superintendent and the heads of each academic department (mathematics, natural philosophy, engineering, French, as well as the professor of chemistry, and the teacher of drawing) and the chief instructors of infantry and artillery tactics. The superintendent was president of the board, but the "head of the academic board" was the senior academic professor. The duties of the board included design of a complete course of instruction, defining exact subjects, apportionment of time to each subject, texts, changes to curriculum, cadet examination, cadet merit standing, and decisions on whether or not cadets continued within their class, repeated a term or year, or were dismissed. The board also recommended the branch of service for each cadet upon graduation.

Any decision regarding the curriculum of the academy had to pass this board by majority vote before being sent to the secretary of war for approval. The secretary normally sought counsel of the chief of engineers and the head of the Fortifications Board. At no time were the general officers in charge of the geographic divisions or those in the general staff bureaus officially consulted in decisions regarding the academy. Therefore, the board set the intellectual tone for the academy and to some extent for the army.

The annual Board of Visitors constituted one way that external forces could influence academy affairs. The practice of using a Board of Visitors to observe and report on colleges and academies was common in the United States at this time.[36] They functioned as public inspectors and could raise issues of concern. Their reports were sent to the secretary of war and made available to Congress and were published openly. However, their observations and recommendations carried only as much weight as they affected thinking of the secretary of war or members of Congress, who ultimately held the purse strings and could insist on change.

The Board of Visitors reports did not alter the fact that during these formative years, the head of the Fortifications Board, Simon Bernard, held the greatest external influence over academic reforms at West Point. He maintained a highly influential relationship with the chiefs of engineers and the secretaries of war, who normally endorsed his reports. In 1820, when leaders in Washington were considering setting up separate schools for infantry, artillery, and engineer officers, Bernard reaffirmed the 1812 decision that all cadets train to a high "elementary" standard in all arms, best achieved in one

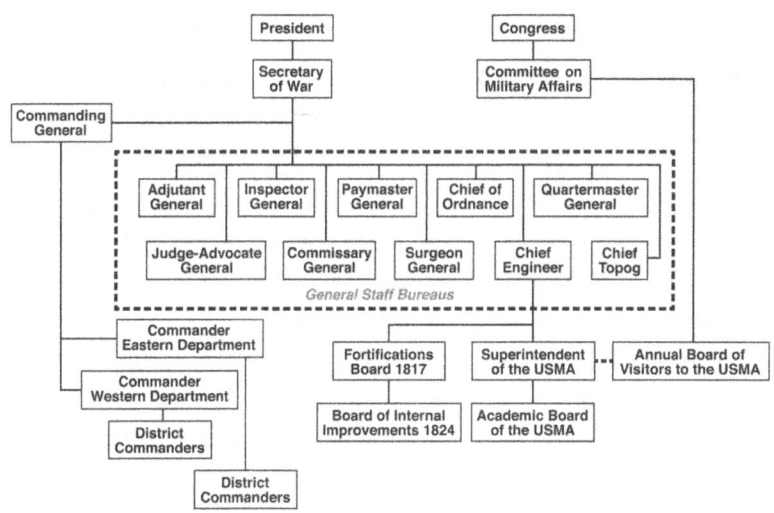

FIGURE 2. War Department staff bureaus and key relationships. Based on *Annals of Congress*, 15th Congress, 1st Session, 268, 273, 289–90, 293, 350, 1568–69, 1687–92.

institution. Bernard felt that this would be critical in an expansible army, where officers would be continually responsible for leading and administering the needs of both infantry and artillery soldiers while at the same time developing military infrastructure. Bernard also saw the all-arms training as essential for the running of a good general staff, whose bureaus were becoming increasingly more reliant on officers borrowed from the line units to fulfill quartermaster, ordnance, commissary, and pay functions.[37]

Under the supervision of Thayer, with oversight from Bernard and Chief Engineer Joseph Swift, the Academic Board changed the curriculum to ensure that all cadets received an adequate level of instruction in infantry and artillery tactics as well as civil and military engineering, ordnance, logistics, and strategy. Secretary of War Calhoun was satisfied. He had become convinced that the main purpose of the academy was to facilitate the "free diffusion of military science" among the regular army and the militia as a way to prepare them for war.[38]

Cadet Selection and the Thayer System

The secretary of war controlled selection to the military academy until legislation in the 1830s placed selection in the hands of Congress, granting representation from each electoral district. Men of political influence had to recommend candidates. This did not prevent indigent youths from securing positions, because part of the mandate of the academy was to educate those who had promise but not means. Once selected, candidates made their way to West Point via New York City or Albany, then upon the Hudson by sail or steam. The only approach to West Point was by way of the river. No roads penetrated the forests and mountains into the point until the 1840s. Once there, the candidates underwent examination to demonstrate and explain some simple arithmetic, such as "the rules of reduction, single and compound proportions, and vulgar and decimal fractions" and to "read distinctly" a sentence from a book and "write legibly" a dictated passage.[39] Unlike entrance requirements for every other American college, candidates did not have to demonstrate knowledge of Latin or ancient Greek or more rigorous knowledge of history.[40]

The entrance exam exposed problems regarding the Jeffersonian ideal of meritocracy and common practices. Until 1816, the secretary of war gave preference of appointment to sons of men who had rendered good service to the republic but who could not afford any other source of education. Second in preference were young men well connected to political leaders. Jefferson himself solicited admission for Nicholas Trist, a family friend, and Andrew Jackson lobbied for admission of his nephews.[41] After his arrival as superintendent, Thayer became conscious that the academy was an attractive place for politicians and members of state elites to send young relatives, regardless of merit, and avoid the costs of expensive education.[42] To impose a rigorous entrance examination would have reinforced this trend; it would have advantaged the sons of the rich and well connected, who could afford preliminary preparatory schooling.[43]

Therefore, Thayer accepted low academic entrance requirements so that the poor might have the opportunity to gain admission, even though this meant that the vast majority passed the initial exam. To ensure that graduates of the academy would be of uniform high quality, however, Thayer raised the academic standards within the curriculum, graduating only 40 to 50 percent of the entering class, with most academic dismissals coming from failures in mathematics or science, wherein standards of evaluation were objective.[44]

That Americans dedicated their singular federal academy to science was not a natural occurrence. In this era, studies in classical languages still characterized postsecondary education, which almost everyone perceived as being exclusively "for the higher classes."[45] All other American colleges maintained a liberal arts curriculum and carried on traditions established at Oxford and Cambridge. Only Norwich University (established in 1819 by Alden Partridge) and the Rensselaer School (1824) eventually offered engineering courses and not until well after West Point had proven the worth of such studies.[46] Further distinguishing West Point was its heavy emphasis on mathematics when almost all other American colleges in 1820 provided only limited instruction in that subject, believing in the English adage that "neither the *mathematics* nor the *physical sciences* are well adapted to develop the faculties of youth."[47] A liberal arts emphasis also prevailed in the Royal Military College at Sandhurst, England, and at the Infantry and Cavalry School at Saint Cyr in France.

The single closest equivalent to West Point in 1820 was France's École Polytechnique.[48] That the War Department chose to emulate this institute rather than conventional colleges was anything but coincidental. The choice marked a clean break with English ways and with aristocratic (and Federalist) structures. The English and aristocratic trappings of America's colonial past were, in Jeffersonian America, expunged as much by deliberate processes as by natural tendency.

Thayer transplanted several important academic practices from the École Polytechnique. One included the division of cadets into classes and sections. The four classes at the academy reflected the year in which the cadet was attending—the Fourth Class being all new first-year cadets, the Third Class consisting of cadets in their second year of attendance, the Second Class having all the third-year cadets, and the First Class being the senior cadets in their graduating year. Thayer subdivided each class into sections of ten to twenty cadets according to the general level of understanding of a subject. Therefore, within any given class there were three, four, or five sections, each under their own instructor. The first section consisted of the brightest and most able cadets of that subject; the last section consisted of the least able. The lower sections were required to muscle their way through the mandatory subject material during the term. The faculty expected the higher sections to achieve this more rapidly and progress to advanced studies in the given topic. Cadets could move up or down the sections within that subject by demonstrating their ability during normal recitations or during exams in January or June.

The pedagogic method remained the same for all: study and recitation. Some professors employed the lecture method with the entire class when texts did not adequately cover certain material. However, most of the learning occurred through recitation. Cadets received a specific portion of a text to know and a specific theorem and problems to solve. Each retired to his barrack room to study the text during the morning and evening study sessions. Using notebooks, cadets transcribed and put to memory specific definitions and theorems and worked out mathematical problems. If a cadet did not finish the prescribed material by the "extinguish lights" tattoo, he might attempt to cheat the periodic sentry check by using candlelight under cover of a blanket or wake early, before the reveille tattoo, and finish his work.[49] The following morning, the professor would call upon a cadet to stand and recite a specific definition or theorem, or he would call forward three cadets at a time to work out problems on chalkboards. Once completed, each cadet would recite the appropriate theorem and explain the formulas used to derive the answers. The others watched and prepared for their turn. In any three-hour recitation session, each cadet could expect to be thoroughly examined.

The "Thayer system" advantaged cadets possessing academic ability. It involved a daily numeric evaluation of two distinct areas, "study" and "general conduct," which together would induce "proper habits." The military faculty evaluated each cadet's "conduct" by penalizing him points for breaches of the academy regulations. Section instructors graded cadet performance in "study" during each recitation, using a point value between 0 and 3 to indicate cadet knowledge of the material:

3.0 perfect
2.5 good
2.0 indifferent
1.5 bad
1.0 very imperfect
0.0 complete failure

Instructors combined recitation points and subtracted points for misconduct every day of the four years that a cadet spent at the academy. They passed weekly totals to the superintendent on Saturday afternoons. In this manner, the faculty kept a running tally on every cadet's discipline and his aptitude or deficiency in each subject. These, combined with exam scores, gave a formal numeric ranking for the cadets of each class and allowed the Academic Board to distribute cadets between sections.[50] As such, this system not only regulated behavior but produced an order of merit in the corps of cadets.

The Thayer system reflected a Jeffersonian ideal of identifying and promoting talent outside of the boundaries imposed by social class or financial status. The superintendent forbade cadets to receive money from their families, so that all cadets, indigent and affluent, had to learn to manage their pay accounts with the exactness that would later be required of an officer managing federal funds on general staff or engineering duties. Financial responsibility was important. Officers—especially engineers, ordnance, and quartermaster officers—had to be trusted with considerable sums of money in building public works and in administering civilian contracts. The policy also enforced financial equality at the academy. One secretary of war stated that Thayer had taken such measures to "foster a feeling of self-reliance and independence destructive of false pride and of all exclusive and aristocratic pretensions."[51] The system offered young men of poor background the opportunity to gain recognition by superiors and peers in a highly structured curriculum where the faculty measured individual performance against "objective" standards, allowing impersonal comparison between cadets.

No other school in the United States followed this system of numeric evaluation of mental habit and conduct, at least until the establishment of state military schools such as the Virginia Military Institute (1839) and the Citadel (1842).[52] However, no other school was dedicated to turning out young men of something akin to a standard product, a scientific officer. In an age when state and sectional identification blended with ideas of personal honor to create general elements in the emerging American character, this effort to homogenize cadets was unique. All civilian universities of this era reinforced the innate American sense of individualism. The practice of offering elective studies, for instance, further reinforced this and spread rapidly throughout colleges in the antebellum period, but not at West Point.[53]

Scientific Curriculum

The academy's curriculum avoided the philosophical and the romantic. The faculty gave no room to hagiographic storytelling or to the metaphysics or dialectics that were emerging contemporaneously in Europe.[54] This was observed by leading American educator George Ticknor, senior professor of French, Spanish languages and literature, and belles lettres at Harvard from 1819 to 1836. Ticknor was a pioneering figure in American education and a good friend of Sylvanus Thayer. He was on the Board of Visitors in 1826, when he extolled the methods and curriculum of West Point. This was something, considering that while visiting Europe previously Ticknor had been exposed to curricula and ways of thinking that were in stark contrast to the method of the academy. At the University of Göttingen particularly, he had witnessed three systems of metaphysics: Kant's, Fichte's, and Schelling's. He observed that there was no dominant system and noted that students were free to choose: "A young man at the university commonly gets this freedom by hearing three or four different professors expound and defend as many different systems." Professors expected students to form their own synthetical personal view. German academics, Ticknor claimed, were eclectics: "The worthiest object of metaphysical studies is to excite and enlarge the faculties, and form deep and thorough thinkers."[55]

Ticknor also observed how different German metaphysics were from the English desire that things "serve some *practical* purpose." He remarked that in Germany there was a separation between men of science and society, and he opined that "from this separation . . . comes, I think, the theoretical nature of German literature in general, and of German metaphysics in particular."[56] Ticknor was observing the fundamental differences that were occurring between German and American military theorists and writers. The Germans were engaging at this time in the practice of dialectical reasoning. The work of the most preeminent theorist of the era, Carl von Clausewitz's *On War* (still being written when Thayer was at West Point), reads like a University of Göttingen debate on the philosophical underpinning of war. Its publishing in 1830 made little stir in the English-speaking world, where preference for clear reasoning and empiricism trumped all effort to promote abstractions. At the military academy, cadets received no instruction in metaphysics "nor hairsplitting argument[s] on the law." Instead, they were to acquire knowledge that was "rigidly demonstrative" and "positively practical."[57]

By 1820 Americans had long established a preference for Enlightenment reasoning, and differences between Jeffersonian and earlier colonial or Federalist thinking of war were marked. Secretary Calhoun and his staff disparaged metaphysics and ideas regarding strict dependency on personality in war. "Genius," Calhoun stated, "may command, but it cannot . . . organize and discipline an army, and give it that military tone and habit . . . to perform the most complex evolutions. . . . Those qualities . . . can only be acquired by instruction."[58] West Point was indispensable in delivering empirical knowledge, absent of metaphysical obfuscations or hagiographic dependency, about proper war preparations because this "form[ed] the basis, in regard to science, on which the military establishment rests."[59]

Alternatives to this way of thinking existed. The obvious at the time was to revert to a military system entirely dependent on amateurs in states' militias. Westerners particularly favored this option. Tennessee representative Newton Cannon in 1820 began the first of what would become a tradition of West Point bashing, claiming that the academy only served an eastern aristocracy and should be abolished, with the money saved going to the militia. However, his 6 February 1821 resolution to Congress challenging the constitutionality of the academy and his motion on 16 February for its abolition were defeated.[60]

Another alternative to granting military command to a small professional officer corps would have been to accept contemporary English military methods where military science was subordinated to aristocratic ideals of war. The English still maintained that war was the realm of the sovereign and that command in war was best invested by His Majesty to men of the social elite who possessed military experience and natural martial genius. Such men would know how to use those skilled in military science without having to possess such knowledge themselves.[61] Many English military leaders felt that technical knowledge was available to the learned amateur and that any officer through self-study could grasp such trade knowledge and apply it by keen mind and energetic spirit.[62] In this Wellingtonian perception of war, theory was not as important as experience and energy, and it was irrelevant without the guiding hand of sublime military genius.[63] Woolwich had existed since 1741, but status as tradesmen held its graduates captive; they were "good officers of Artillery and perfect Engineers" but not military commanders.[64]

From the 1770s until well into the 1830s, the English remained skeptical about the very idea of a science of war. British writer and instructor C. Malorti de Martemont put it thus in 1806: "I cannot . . . pretend that an art as that of war . . . ever will . . . become a science, . . . that it will no longer . . . excite an ambition to possess it, and that consequently men of genius will prefer directing their attention to objects of more general utility."[65] Among the considerable number of self-educated and professionally minded British officers, private appreciation for the science of war never grew to the point where it could openly challenge the consensus that military wisdom resided in the "practical sagacity" of "social betters."[66]

The instruction of the "complete course" of studies at the U.S. Military Academy emphasized a different relationship between command personality and military science. It rejected the idea that victory in war depended upon the innate decision-making qualities of those who were "social betters" by birth right. Instead, it sought to provide officers drawn from various social backgrounds from each state of the union a common foundation in military science upon which "sublime" qualities such as *coup d'oeil* or genius might become more discerning. As a foundational institution, the academy subordinated the role of individual sublime talent to a collective appreciation for all of the branches of military science. It was felt that scientific knowledge would enable a commander to choose the best tactical application of each of the branches of military science to suit the local circumstance and in doing so genius— if it were present at all in the character of the commander—might be displayed.

Calhoun supported the academy's role in producing learned professionals. Graduate general officers were serving him well, particularly Alexander Macomb and Joseph Swift. He envisioned that West Point graduates would supply the commanders of the future. They would not replace the indispensable state-appointed commanders of state militias, who were indeed the American equivalent of England's "social betters." However, Calhoun realized that West Pointers would occupy some key command and most key staff positions in the army and that their mathematics and scientific knowledge were indispensable.

Those officers and professors who were committed to reforming West Point, and who were responsible for producing officers who could fulfill staff functions and become commanding generals, realized the fundamental importance of mathematics. In fact, they held mathematics to be the most important component of the new curriculum, certainly more so than the study of classical languages and literature. They also felt that mathematics was more egalitarian than the study of ancient Greek and Latin. Their decision to reject classics in favor of mathematics did not go unchallenged. The army's inspector general, Brig. Gen. John Wool, stated as much when he reviewed the proposed curriculum. In 1819 he wrote to Secretary Calhoun to criticize the emphasis on arithmetic and engineering at the expense of study in history, geography, and languages. Great victories, Wool claimed, "were not achieved by the 'rule and compass' or the 'measurement of angles.' They were the product of enlarged minds, highly cultivated and improved by a constant and accurate survey of human events."[67]

Aware of such sentiments, Calhoun seriously considered splitting the instruction at the academy into two courses: the first for infantry, artillery, engineer, and topographical officers combined, and the second for artillery and engineer cadets to learn advanced mathematics and fortifications theory. A Board of Visitors to the academy made the same proposal in 1821.[68] Both proposals were rejected because of the additional expenses required, because of the possible dilution of the comprehensiveness achieved under the Thayer system, and because Simon Bernard did not agree.[69] However, criticism of the preponderance of mathematics in the curriculum was to recur periodically over the course of the next century.[70] All challenges were unsuccessful, and mathematics remained the most important academic subject of instruction at West Point.

The challenges were not from external sources alone. Even faculty members thought that the arts warranted greater emphasis. In appendixes attached to the proposed 1819 *Regulations* of the academy, the Academic Board signed a list of twenty-four propositions for consideration by the chief of engineers and the secretary of war. Proposition 2 called for a professor of geography, history, and civil law and for a professor of languages, oratory, and belles lettres. In formal response to this, Simon Bernard provided these qualifying remarks: "changes . . . should be confined to what is strictly necessary; and not to hazard the introduction of accessories. . . . Geography, history, jurisprudence, the languages, belles-lettres, are considered as among the number of these accessories."[71]

Bernard's ideas regarding the curriculum carried greater weight, and the secretary of war rejected the propositions. To Bernard, the Third System requirements and his, Crozet's and Thayer's unwavering conviction in the Polytechnique model combined to ensure that all cadets would acquire *mathematical minds*, regardless of future branch assignments. Faith in science and its utilitarian purposes trumped desires to instruct liberal arts subjects. Thayer supported Bernard's way of thinking, but not because he felt these "accessories" to be unimportant. He believed that officers would have considerable opportunity to learn "general history" and "ethics" and other subjects after graduation, but also that proper instruction of natural philosophy, engineering, and topography could only happen at West Point.[72]

Calhoun approved the recommendations of his engineers because he understood that a federal system of defense depended on scientific officers. Fear of a repeat of the invasions of 1813 and 1814 motivated thinking of federal regulation and coordination of all defensive efforts. The agents of this coordinated system were to be the West

Point graduates. The academy needed to ensure that these graduates formed a corps of professional officers dedicated to service to the union. Their course of instruction would give them a common and firm grounding in mathematics, followed by knowledge and skill in artillery application, logistical science, engineering, and the science of marches. Little room was left for "accessories" that were also considered the staple of an aristocracy.

Royal Navy incursions into the Chesapeake and Potomac in 1813 and 1814 left residual fear among politicians and peoples of the mid-Atlantic seaboard. The havoc caused, together with the inefficiency of the existing military establishment to deal with threats, prompted considerable impetus for reform in the period 1815 to 1825. Presidents James Madison and James Monroe guided the reform, leaving the details to Secretary of War John Calhoun and engineers of the regular army, including French Napoleonic officers (and professors) brought to the United States for this very purpose. Organizational reform included the creation of staff bureaus in Washington under secretarial control and an expansible regular army. Formulation of a coherent defense plan based on fortified maritime frontiers and a system of national mobilization of militia forces gave the United States its first national military policy. Adoption of the French Napoleonic model of instruction and doctrine began a process of regulation in the army that was to last until the Civil War. With Thayer as superintendent, the U.S. Military Academy underwent fundamental reforms to bring it in line with the vision of an expansible army ready to implement national policy. In this, Americans were following a French model of military professionalization. Cadets underwent a process of military socialization to acquire the "habits" that Thayer felt were needed of professional officers. The faculty instituted a method of assessment to promote and reward proper study habits and good conduct. The process of socialization changed very little between 1820 and 1860 and was an important component of cultivating military science as a hallmark of the regular officer corps.

The plain of West Point was in this way becoming a great leveler, where talents of mind were recognized above status of birth. The growth in percentage of graduates within the regular officer corps demonstrated the success of the Thayer system. In 1820 graduates constituted only 18 percent of the 540 regular officers. Ten years later, they represented 64 percent of all officers. By 1860, 76 percent of 1,108 officers were graduates. Resignation rates during this period averaged 4 percent, and promotion was slow. The average career for the West Point graduate during the antebellum era was just over twenty years.[73] In all of this time, the army never posed a threat to civil authority in Washington and remained (with the exception of several high-profile cases) politically unaligned.[74] Under Thayer, West Point became an institution dedicated to the creation of a small professional officer corps not beholden to any specific political patronage. It was expert in military science and committed above all to improvements of the nation's defensive infrastructure and to preparations for war.

3

West Point's Scientific Curriculum

The Thayer system subjected cadets to a rigorous and rather un-American daily routine that reinforced the rigor of the "complete course" of military science delivered in the curriculum. The subjects and program of study were the same for every cadet. The Academic Board allowed neither electives nor deviations. The routine and curriculum instilled basic skill and knowledge in all branches of military activity and created officers who could perform any number of line or staff functions in the small and highly dispersed army or in the vastly expanded military establishment envisioned during mobilization. In this respect, West Point was unlike any other military school. The academy's curriculum was the most well rounded of any offered in Europe or North America. Henry Barnard wrote in 1862 that it allowed the cadet "to acquire that profound knowledge of the science and material of nature, which should fit them for the complicated art of war; to defend and attack cities; to bridge rivers; to make roads; to provide armaments; to arrange munitions; to understand the topography of countries; and to foresee and provide all the resources necessary for national defence. This was the object of the Military academy."[1]

Cadet Routine

For their four years at West Point, the cadets awoke just before sunrise each morning (save Sundays) to the sound of the academy drums.

SCIENTIFIC CURRICULUM

They quickly assembled outside of their barracks for mandatory roll call, drill (or study), inspection, and breakfast. At 8 a.m. each class and section then began a three-hour recitation in the primary subject of the term, followed by two hours of study in the cadets' barrack rooms (see 1, 2, and 3 in appendix of tables for cadet daily routine). Academy drummers beat the call for dinner parade at 1 p.m. Afternoons were spent in the study and recitations of the second course of the term from 2 until 4 p.m., and at military exercises from 4 until 6 p.m. Here junior cadets focused on foot drill and the manual of arms and cannon crew drill. Third Class cadets exercised in the "school of the company" (60 to 100 cadets conducting evolutions and firings), artillery battery drills, and the duties of a corporal. Second Class cadets concentrated on the "school of the battalion" (evolutions and firings of a group of 400 to 1,000 soldiers), artillery battle tactics, and the duties of a sergeant. First Class cadets learned orders of battle of the line (an army consisting of multiple battalions grouped into brigades of three to four battalions each), marches, advanced artillery tactics, duties of the officer, and use of the sword. Within the four years, all cadets had learned and practiced how to organize, train, and command up to a battalion of infantry or a regiment of artillery. The regular army contained no cavalry units until the establishment of the first U.S. dragoon regiment in 1833, and it was only in the mid-1830s that the Academic Board added cavalry tactics and horse artillery drill to this instruction.

Upon completion of military exercises, the academy drummers beat supper parade at 6 p.m., and after the requisite thirty minutes to dine, cadets marched back to the dorm rooms where they were required to stay and study. The drum tattoo for evening roll call sounded at 9:30 p.m., followed by the drumming of "lights extinguished" at 10 p.m. This daily routine did not vary between September 1 and June 30, except that classroom recitations in Tactics replaced most out-of-doors military exercises from November to March. Weekends started with Saturday morning recitations, but the cadets were granted a free period between dinner and 4 p.m., followed by one hour of cleaning of kit. Cadets were free on Saturday evenings, and they started Sunday mornings with a 9 a.m. inspection followed by a mandatory church parade. They recommenced the routine of study periods and recitations after dinner on Sunday.

One can imagine the numbing nature of this routine, particularly for those youths who had come from relatively leisurely lifestyles. The regulations governing cadet conduct at West Point prohibited drinking, fighting, playing cards, or games of sport. However, cadet memoirs and disciplinary records attest to the fact that they managed to continue all such activities clandestinely during their limited free time. Regulations forbade cadets to leave post, possess or receive money from outside sources, or subscribe to civilian newspapers or journals without approval of Major Thayer. While cadets were allowed to make withdrawals from the academy library, this could occur on Saturday afternoons only, restricted to one book at a time, and, until the late 1830s, the book had to relate directly to their subjects of study that term.[2] The library carried no works of popular literature or poetry until the 1830s, and regulations prohibited ownership of a novel because the Academic Board considered these to be "accessories" that diverted attention from the primary academic subjects.[3] The Academic Board did not prohibit cadets from forming extracurricular literary, debate, or theatrical societies, as approved by the superintendent. Cadet reminiscences of this period are full of stories of activity in such clubs, and many of the club documents still exist. However, the superintendent allowed cadets to participate in these only so long as their academic marks remained high.[4]

Upon completion of the June exams, the corps of cadets left barracks to encamp on the West Point plain only a few hundred yards from their dorms until the end of August. The military camp had its own routine. Academy drums woke the cadets at dawn for drill and military instruction before breakfast. Each morning and afternoon cadets practiced platoon, company, and battalion battlefield evolutions and gun, detachment, troop, and battery artillery drill. In the 1830s, cavalry and horse artillery drill were added, as well as rudimentary field fortifications and training in pontoon and expedient bridge building. The purpose of the encampment was twofold. First, it allowed cadets to learn castrametation (how to encamp properly) and all of the necessary practices of units living in the field as they would on active campaign service, enduring variances in weather and practicing the personal economy and collective logistics needed to sustain soldiers on operations. The second aim was to ensure that graduates left the academy with knowledge of everything required of the soldiers they would command. Experiencing the life of a soldier for three summer encampments made cadets thoroughly acquainted with soldier routine. No other officer training institution practiced this egalitarian approach to forming a professional officer corps.[5]

The Curriculum ca. 1825

Instruction in Tactics was not limited to summer encampment. During the academic year, starting on September 1 and ending on June 30, it constituted the largest component of instruction, with 2,064 hours of recitation and practice in theory and drill. In academic instruction, mathematics and French were of primary importance and were the only subjects taught during the first two years, occupying 1,548 hours and 1,206 hours, respectively. Mathematics was essential for three reasons. Cadets needed elementary arithmetic to understand tactics, simply because marches, encampments, and combat required calculation of rates of movement and fire in columns and lines. They required advanced mathematical knowledge as prerequisite to learning subjects taught in the third and fourth years of study. Finally, knowledge of some mathematics was required after graduation for work in staff bureaus or in secondment to one of these bureaus from the line. In 1820 one-fifth of army officers were engaged in such staff activity.

Instruction in French allowed cadets to read and translate the best texts on military science and tactics, most of them brought over from France. English military texts existed at the time, although various trade embargos made them less accessible. However, Chief

Engineer General Joseph Swift and the West Point faculty found that these did not render systematic and comprehensible instruction in all branches of military science and contained "vacancies of reasoning."[6] Swift remarked that "the sciences that are taught in the continental schools are admitted to be more systematic than in England," and the 1823 Board of Visitors reported that "the French under Napoleon, were something like a century ahead of every other nation in the art and science of War."[7] These sentiments reflected a general trend in American society favoring continental scientific texts. Many of America's most prominent schools were increasing French language instruction to afford students opportunity to access French works.[8] Approximately 570 of the 940 books held in the academy library in 1822 were in French, and this proportion would continue until the 1840s.[9]

After acquiring a strong foundation in mathematics and French, cadets were ready to advance to more involved study in their final two years. The primary subjects were natural philosophy (with 1,032 hours of dedicated instruction), engineering and military science (also with 1,032 hours), drawing (776 hours), geography, history, ethics, and law (which shared 516 hours), and chemistry and mineralogy (granted 130 hours each). These allocations changed remarkably little between 1820 and 1854, when the academy initiated a five-year program.

Fourth and Third Class Instruction

A cadet's first terms of study in French consisted of selected chapters from a "reader."[10] French grammar was not the focus so much as translation and understanding. Cadets were required to read aloud particular passages, practicing correct oral pronunciation and translating the whole into English on the blackboard. The assigned readers exposed cadets to a broad vocabulary, and the moral constituents of each story provided words, phrases, and ideas appropriate for a gentleman's conduct. They also contained vignettes from literature and history, as well as lessons in geography or moral instruction.[11]

However, the principal subject during a cadet's first terms was mathematics. Here he studied basic algebra, geometry, and trigonometry. Mathematical theorems and definitions were important, and cadets carefully transcribed these into notebooks and then memorized them. Mathematical instruction was progressive.[12] Cadets learned that the components of geometric measurements—atoms, lines, angles, and planes—were the means by which they might precisely measure distances or dimensions on the earth's surface and that this had direct applicability in planning to attack or defend a position. From the simplest constructs, cadets progressed to understanding the importance of perpendicular lines, obtuse and oblique angles, and parallel lines, all of which would allow them to understand tactics and strategy. Likewise, they came to appreciate the characteristics of geometric figures (squares, rhombus, trapezoid, polygons, and triangles) and their importance to fortifications design, for as Thomas Paine had stated, "A place that cannot be enclosed in a polygon cannot be fortified on any principles of fortification."[13] Cadets also learned algebra and logarithms, thus providing them with a method of computation and allowing them to progress to plane trigonometry, by which, with certain data, a cadet could determine the unknown parts of triangles. With this knowledge an officer could calculate ranges of engagements and angles of fire and the appropriate lengths, widths, and heights for fortifications—calculations essential to the attack and defense of places and in setting out camps and battle formations. Existing cadet notebooks show calculations for these purposes, illustrated with sketches of ships, soldiers, fortifications, and cannon.[14]

Daily instruction in tactics reinforced cadet appreciation for the academic instruction. Von Steuben's *Regulations* and Louis de Tousard's *American Artillerist's Companion* (Philadelphia, 1809)[15] were standard texts until replaced by Winfield Scott's *Regulations* and his Americanized version of the 1791 French infantry drill manual, *Infantry Tactics; or, Rules for the Exercise and Manoeuvres of the United States Infantry*. During the fall and winter months, tactics involved theoretical recitations, wherein cadets memorized and recited portions of these texts regarding soldier and company evolutions and drills, as well as readings on surveying, reconnoitering, and castrametation.

In the original West Point curriculum, a cadet could master the mathematics, French, and minor tactics in ten months. In Thayer's curriculum, cadets had subjects that took considerably longer to grasp, such as analytical trigonometry, calculus, and spherical projections. Claudius Crozet introduced descriptive geometry to Third Class cadets in 1817. Other American colleges neglected this subject—long considered fundamental in most French public schools—until the 1830s. Cadets used it to calculate dimensions of spheres and relationships of planes to spheres in three dimensions, an essential skill for an officer planning permanent fortifications or civil engineering works. Without skill in descriptive geometry, an engineer would have to build a small-scale model of the desired construction and project from this the dimensions and angles necessary. However, models were useless for adjusting measurements to compensate for variations in actual terrain and hydrology. Descriptive geometry substituted mathematical models for actual models and granted much greater accuracy in projections of dimensions and angles. In France, the use of descriptive geometry distinguished competent from average civil engineers. Professors at the military academy used it as the distinguishing subject for determining which cadets possessed the aptitude to stay in the top sections of their class and possibly become engineer or artillery officers and which would remain in the bottom sections of the class and receive an infantry commission.

After cadets learned descriptive geometry, they progressed to spherical trigonometry, a subject related to navigation and map making. Solutions to all problems of unknown latitude, longitude, variation, etc., were attainable using trigonometrical calculation. It was also essential to an officer seeking to survey, compensating for curvature of the earth's spherical surface. Descriptive geometry and spherical trigonometry consumed most of the cadet's second year of instruction, but time was reserved for professor Thomas Gimbrede to teach elements of landscape and perspective drawing.

In second year French, cadets translated Voltaire's *History of Charles XII*, which served the dual purpose of language instruction and historical analysis of a commander in war. The choice of Voltaire's work is indicative of how the structure of studies at the academy provided mutual reinforcement of one subject to another. Voltaire promoted empiricism, warning his reader to be skeptical of emotional historical accounts and to "distrust whatever is marvelous."[16] However, it was Voltaire's thoughts of war that dominated most of this French language study. Voltaire exalted Swedish discipline and the employment of military science by its first masters, Gustavus Adolphus and Charles XII, who knew war as a matter of planned marches, raised camps, fortified redoubts and trenches, and strict discipline.[17] Charles's Swedish army had become masterful in such applications. However, cadets learned that although Charles's victories "bordered on the marvelous," the king's ego proved fatal: "he carried all the virtues of heroes to an excess at which they are as dangerous as their opposite vices." Charles's wars ended with the near destruction of Sweden after nine years of offensive campaigning that were more important to the ego of the sovereign than to the welfare of the state.[18] The drama of Voltaire's account rose until the final part of the history when a cannonball struck down Charles at an exposed angle in the trenches while the king was surveying the progress of the siege of Frederickshall in December 1718. Voltaire ended with a warning that young officers should heed: that "a pacific and happy government is preferable to so much glory."[19]

Second Class Instruction

In their third year, cadets learned natural philosophy, chemistry, drawing, and the science of artillery.[20] The course in natural philosophy required 1,032 hours of instruction in statics, hydrostatics, hydrodynamics, hydraulics, pneumatics, optics, and practical astronomy. Instruction in chemistry included pyrotechnics and basic mineralogy.[21] Drawing classes included the study of landscape, the art of shading, geometrical figures, sketches from nature, and elements of topographical sketching and map making.

The military application of all of these subjects is not obvious, particularly to those predisposed to seeing military matters as strictly things important on a battlefield. In and of themselves, these courses may be construed as delivering mere technical knowledge, important only to those cadets destined to become engineers. Samuel Huntington and other late twentieth-century commentators have made this argument and condemned the academy for serving narrow purposes. However, in 1820, and for the next three decades, the majority of graduates became officers of the artillery, engineer, and ordnance branches. The Academic Board recommended branch assignments only at graduation with only a minority of cadets going to the infantry and none to the cavalry before the mid-1830s. The Board understood that all cadets, regardless of branch assignment, were eligible to serve in direct command of garrison artillery or be attached to technical staffs. These factors made technical instruction entirely relevant to all cadets, and they substantiate why this continued throughout the antebellum period.

All of these subjects were of fundamental importance to artillery officers, and most pertained to engineers as well. The duties attached to artillery service in the antebellum era were not simply to fire cannon in battle but also included "the direction and establishing of manufactures and foundries," the "safekeeping of arsenals," the construction of wagons and carriages, the making of munitions, the establishment of depots, and the "preservation of stores."[22] The War Department had widely dispersed the regular army across the expanding country and granted them a mandate to prepare for war. This involved all aspects of readiness, such as the preparation of American industry for the manufacture of war equipment, the casting of cannon and musket and all types of shot, experimentation in ammunition, weaponry, and accoutrements, and the planning and construction of infrastructure dedicated to military preparation and defense, such as armories and foundries.[23] All regular officers, even comparatively low-ranking officers, were subject to employment overseeing public works, surveys and expeditions, and other technical assignments of the general staff departments. It is in this strategic context that the third year program of the military academy gains meaning beyond its mere technical content.

The tactical instruction of the Second Class also related directly to this academic curriculum. Instruction concentrated on the theory and exercise of seacoast batteries and gunnery pertaining to howitzers, mortars, and heavy and light cannon. Specifically cadets learned pyrotechnics, the manufacture of gunpowder, the casting of cannon barrels and projectiles, and the construction of various gun carriages and caissons. They also learned how to create breaches in sieges and how to construct bridges and ferries. These subjects complemented instruction in natural philosophy, machinery, and chemistry.

The Second Class studied and recited from H. Lallemand's *Treatise on Artillery*, published in 1820 as translated from French by Colonel Renwick (upon commission of Chief Engineer General Swift and by encouragement of Major Thayer). Historians will probably never know exactly how much of Lallemand's work was covered; however, the number of hours of recitation available for tactics each year (over 500) would suggest that cadets read a fair portion of the text. It is useful, therefore, to examine it, and the principal tactics text of the First Class—Gay de Vernon's *Treatise on the Science of War and Fortifications*—for purposes of demonstrating the key components of nineteenth-century military science taught at West Point. These texts were not merely branch booklets outlining the drills and evolutions of battle. They also were encyclopedic renderings of a "complete system" of warfare—mainly Napoleonic in character, and certainly Enlightenment in origin—embraced by the U.S. Army and foundational to the guiding paradigm for the remainder of the century. In this paradigm, military engineering was essential to preparation for war, and artillery was the decisive arm of battle. Knowledge of both branches was the apogee of military expertise and essential to acquiring skill at strategy.

Renwick modified Lallemand's work to suit American needs, a point explained in the introduction. Because European armies contained regular artillery units and military "schools of practice" in all large garrison towns, they had less need for published works on artillery accessible to common citizens. Lack of any equivalent schools in the United States and reliance on mobilized citizen armies demanded something different—a publication that recognized that "all the citizens are, without distinction, called to the service of their country."[24] Therefore, Renwick meant for his version of Lallemand to be comprehensive. It outlined the types and organization of artillery, the theory and tables of artillery fire, maneuvers and evolutions of field batteries, the construction of entrenched batteries, military bridges, army, corps, and divisional artillery on campaigns, the attack and defense of fortifications, military reconnoitering, artillery logistics, and the manufacture of ammunition and guns.[25]

Cadets learned that artillery was a broad art, involving the construction, preservation, and use of "three species" of weapon—the gun, the howitzer, and the mortar—in their variances of caliber, throwing 6 pound, 12, 14, 18, 24, or even 36 pound balls or shells.[26] The firing of artillery involved mathematical calculation using range, trajectory, weight of shot, amount of gunpowder in the charge, and terrain variations to determine the angle of fire of the gun, an activity validating the mathematics instruction of the first two years of study. Yet as important as was cannon fire, Lallemand's text gave equal detail to construction of carriages, wagons, wheels, sledges, mortar beds, and ammunition caissons.[27]

The movement of artillery was of equal concern, and Lallemand paid attention to the role of specialist artisans and pontoniers. Cadets learned about the importance of the construction of pontoon bridges and boat, swing, carriage, trestle, and permanent bridges.[28]

Lallemand also touched upon the strategic elements of the artillery art, the organization of army artillery to support operations, and the management of corps and army artillery reserves.[29] In advocating artillery parks and logistics, Lallemand dismissed the idea of living off the land, because artillery, unlike cavalry and infantry, must draw from depots and magazines that in turn rely on national foundries and armories behind the military frontiers.[30] This point is extremely important. It framed thinking about military campaigning and strategy and emphasized the essential nature of depots and magazines and convoys to all campaign planning.

Cadets learned from this instruction exactly what Third System advocates were suggesting: that proper war preparation connected efforts at national military depots to efforts on military frontiers. Lallemand also emphasized the need for continuous education and practice and experimentation in order to prepare officers to solve problems in war. He felt that an officer needed a sound knowledge of theory and the ability to "repeat carefully all experiments having relation to his profession, and moreover . . . not have lost the habit of thinking, and of reflecting upon questions already discussed."[31]

Lallemand dedicated a portion of his text to army command and control. Of command he cautioned that "success does not depend upon genius alone, but also upon the intelligence with which orders are executed."[32] Therefore, Lallemand suggested having a major-general chief of staff, and personnel available to reconnoiter and transmit orders and help compensate for lack of genius.[33] He also championed the Napoleonic corps structure comprising three divisions of infantry and a reserve of at least a brigade, four regiments of cavalry, and artillery pontoniers to construct and repair bridges. The force would total between 20,000 and 40,000 men.[34] This system was unique to the French and had allowed Napoleon's Grande Armée to operate with an agility and efficiency never before seen. While slow to develop, other armies did eventually move to group brigades and divisions into corps, a proposal made by West Pointers to President Lincoln in 1862.

Beyond tactics and logistics, cadets may have learned elements of strategy from Lallemand's text, which gave definitions and essential descriptions of strategic terms and practices. Lallemand defined strategy as "the science . . . of employing troops in the execution of a campaign; that combines and directs the several operations to a determinate object; that anticipates lines of march, camps, and fields of battle."[35]

Lallemand's strategy depended on knowledge of topography of an enemy's frontier, and he described in detail the art of military reconnoitering.[36] This included understanding topographical features, field sketching, preparation of maps, and the collation of information into a "Descriptive Memoir" which discussed "the properties of the ground in relation to the marches and movements."[37] The production of a good descriptive memoir required knowledge of geometry, geography, mathematics, and the governing principles of military strategy. Lallemand's ideas on reconnoitering were reinforced in the Second Class course in drawing, which covered field sketching and mapping.

Lallemand thought that skill in reconnoitering could help develop "sublime" qualities, and it is here that he introduced the continental notion of military genius, *coup d'œil*, of which he recognized two varieties. The first was "a faculty by means of which an officer takes, in one comprehensive view . . . the topographical and geometrical surface of the country [and] . . . the military properties which it possesses."[38] The second occurred when a commander "penetrates into the views, the object, and the plans of an enemy, . . . inspir[ing] those great thoughts that destroy his combinations and command success." This second variety of *coup d'œil* Lallemand called "a Genius for War."

Lallemand sought in his treatise to convey all that an officer needed to know to acquire the first variety of *coup d'œil*, trusting that it might prevent military disaster if the second variety—genius—failed to emerge. After all, "nature only produces men of such a stamp from time to time." Lallemand remained adamant that nations should count more on skill than on genius "for the fate of their military destinies." He believed strongly in the use of rules and maxims, which he claimed were "deductions from the labours of the man of genius [a Frederick or a Napoleon]" and should guide action in the absence of actual genius. Acquiring military information and knowledge—military science—therefore was essential, because "all things else being equal, the probability is that victory will rest with the disciplined army."[39]

Recitations from Lallemand reinforced the belief that war was a learned vocation and that experience alone could not produce a first class commander. The Napoleonic philosophy held that victory in war came not from a series of random acts of genius and contingency but with the application of a complete military system first learned through study. The subordination of notions of glory and genius to knowledge, rigorously learned and uniformly applied, was the essential principle in West Point military science. Artillery knowledge and skill—not just in firing cannon but also in casting, mounting, moving, and sustaining artillery in support of the nation's military establishment—was a large component of the knowledge base of the professional officer. This was particularly true within the Third System, which was about constructing fortifications designed to protect artillery batteries that would batter an enemy's invasion fleet. Of course, this also required engineering skill—the focus of the last year of study at the academy.

Instruction of the First Class

Some knowledge of military engineering was essential to all regular officers, particularly given the nature of the Third System. Therefore all cadets learned the basics during their last year, along with military strategy and drawing. Those of the upper sections learned the full range of both military and civil engineering, although in the early 1820s the civil component was limited to the theory of construction of bridges, canals, and roads.[40]

The guiding text throughout the year was another French work,

SCIENTIFIC CURRICULUM

Gay de Vernon's *Treatise on the Science of War and Fortifications Composed for the Use of the Imperial Polytechnick School, and Military Schools, and Translated by the War Department for Use of the Military Academy of the United States to which is added a Summary of the Principles and Maxims of Grand Tactics and Operations, Treatise on the Science of War and Fortifications.*[41] It is a curious act of American Enlightenment thinking that allowed the West Point translator, Capt. John Michael O'Connor, to take the original text title, *Traité élémentaire d'art militaire et de fortification*, and substitute the word *science* for *art*. To appreciate fully the military science of antebellum America, one must examine this text, as it was instrumental in shaping American thinking, at least until Professor Dennis Hart Mahan replaced it with his own notes and texts in the mid-1830s. Even then, much of Mahan's work was Napoleonic in character.

That O'Connor embraced the Military Enlightenment is evident in his introduction:

> The march of the human mind is steady and progressive; and like the motions of the heavenly bodies, is guided by a Hand divine. . . . It is to be hoped that War, which has ever kept pace with civilization, and become scientific in proportion as the arts and comforts of man have increased, may be the means in its improved state of liberating that world which barbarous violence enslaved.[42]

O'Connor believed that on this "march of the human mind" knowledge of war had reached a perfect state and could reduce senseless slaughter in battle. He thought that Gay de Vernon's text came close to being a complete cataloging of knowledge of war. To make the text perfect, he added his own appendix on military strategy, derived from the best-known European authors: Lloyd, de Guibert, von Bülow, and Jomini. O'Connor believed that the entire work "contain[ed] an epitome of all that is known in war," giving cadets "the essence of all military truths."[43]

Gay de Vernon had written his *Treatise* to educate all officers.

Part 1 presented the general nature of war, what constituted military knowledge, *coup d'œil*, army organization and the roles of each branch of service, weapons, general principles of orders of battle, castrametation, and military frontiers and topography, as well as the basics in "arithmetic, elementary and practical geometry, and drawing." Part 2 covered field fortifications in all possible variances of terrain. Part 3 was written for artillery and engineer officers, and presented a full course on permanent fortifications, including Vauban's methods and modifications by Louis de Cormontaigne and other French engineers. Gay de Vernon's *Treatise* covered almost the full spectrum of military knowledge and advocated science as the means to unlocking the secrets to general military preparedness. He added that skilled scientific officers could form an excellent general staff and "constitute an inestimable corps from which the sagacity of Generals may draw all the resources productive of victory."[44] To complete this work, O'Connor added an appendix summarizing the principles of the *Plan of Campaign*, *Strategy*, and *Grand Tactics*, giving historical examples.

The first important characteristic about the *Treatise* is that it was not a textbook on tactical drills and evolutions and it was not specific to the infantry, artillery, cavalry, or engineers. It instead provided a doctrine of warfare for all readers. Significant to the United States, it did not cover *la petite guerre*, a deficiency addressed partially by Professor Mahan in the 1840s. Instead, Gay de Vernon concentrated on conventional state versus state war, *la grande guerre*, characterized by the use of fortifications, ships, and armies locked in intense combat wherein artillery was the key to success. The *Treatise* concerned how to prepare for and conduct large-scale operations in a theater of war. From it, cadets could learn higher military thinking, delivered as a coherent whole, dedicated in the main to the preparation and use of fortified national frontiers to conduct both defensive and offensive campaigns. General knowledge of this would provide cadets with necessary context for acquiring more detailed knowledge of each tactical branch of the army.

Three related themes from the *Treatise* had lasting impact on subsequent American military thought. The first was the importance Gay de Vernon placed on permanent and field fortifications. The second was O'Connor's summary of campaign theory based on geometrical reasoning and topography, military reconnaissance, and campaign marches. The third important theme was the insistence that officers acquire knowledge of warfare by rigorous study to reinforce experience and to balance any lack of innate talent or genius.[45]

The primary concern of cadets of the First Class was knowledge of fortifications: the branch of military science upon which the principles and maxims of all the others depended, for "no person can properly understand grand operations, without a previous knowledge of fortifications, which by its truths and reasoning prepares the mind for their reception and comprehension."[46] A connection between fortifications and general campaigning had existed since Vauban.[47] They shared common principles. Cadets learned to first apply fortifications theory: understanding the importance of defensive bastions as mutually supporting positions designed to stop attacks, angular projections that helped to concentrate flanking fire, and siege parallels that served to concentrate fire and as secure points to stockpile stores in preparation to launch attacks. The principles transcended to the use of mutually supporting frontier fortresses to check invasion, the importance of concentrating force upon the flank of an enemy army, and the use of fortified points as bases from which to commence or recommence offensive campaigns.

Individual fortifications, no matter how small, presented obstacles to an enemy and protected and even increased the strength of infantry and artillery therein. Their purpose was to give shelter to defending troops against missiles and "to arrest the march" and bring effective fire upon attackers.[48] All fortifications used wooden or masonry walls and thick earthen ramparts to provide this shelter. They also used angular projections in the rampart (called bastions) to achieve flanking fire upon assailants.[49] A bird's-eye view of any fortification built in the Vauban style shows polygon designs with

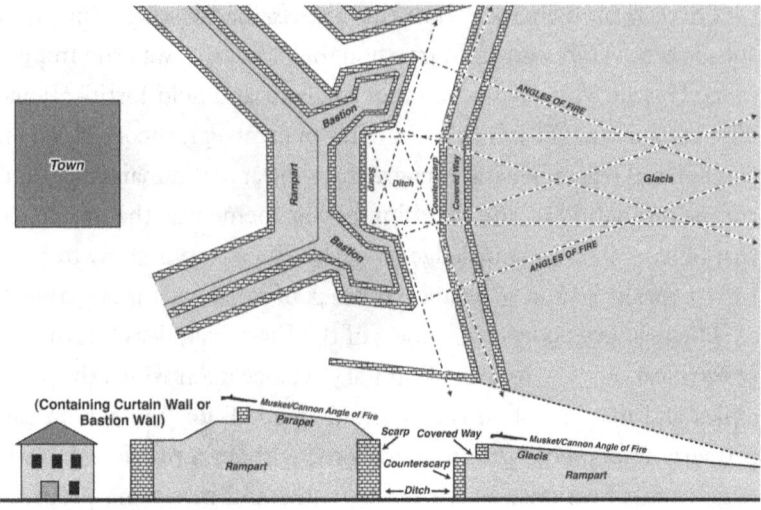

FIGURE 3. Vauban-style fortifications with trace showing concentration of flanking fire. Based on *Treatise on the Science of War and Fortifications Composed for the use of the Imperial Polytechnick School, and Military Schools, and Translated by the War Department for use of the Military Academy of the United States to which is added a Summary of the Principles and Maxims of Grand Tactics and Operations* by Simon François Gay De Vernon, translated by Capt. John O'Connor (New York: J. Seymour, 1817), plate 4.

triangular, quadrilateral, or pentagonal bastions. Engineers transformed terrain in such a way as to allow a defender to produce artillery and musket fire evenly in all directions, but with each angled wall providing fire in mutual support of another wall or bastion, concentrating that fire on an attacker's front and flanks simultaneously. This principle had not changed since first employed in the *trace Italien* in the sixteenth century, and it persisted until after the First World War. The design and construction of Third System fortifications contained mostly Vauban characteristics.

A cross-section of any fortifications of this era shows similar essential characteristics (see fig. 3). Builders removed earth from a ditch and piled it into a rampart at least forty-five feet thick and formed into a basic polygonal design. The exterior side of the rampart was often a masonry scarp—the curtain wall—that held the earth in place. The height of the curtain was optimally thirty feet, the height at which scaling ladders were too heavy and cumbersome to carry. On top of the rampart was an earthen parapet that extended from the top of the curtain wall back for seven to twelve feet, protecting a firing platform that was four and a half feet below the interior height of the parapet. This was believed to be the average height needed for a man with a musket to fire over the parapet with most of his body concealed (such measurements remained constant from the seventeenth century to the end of the nineteenth century). Builders covered the top of the parapet in turf and sloped gently downward toward the top of the curtain wall. In front of the rampart was a broad ditch, on the other side of which was another rampart lower than the first, with a palisade on top and a long sloping front that ran from the palisade down to ground level in a gentle and uniform gradient, exactly the same as the slope of the higher parapet. This was the glacis. It was designed to absorb cannon shot fired at the curtain wall and to ensure that enemy infantry advancing toward the ditch were exposed to the shot from the parapets above them.

When looking at the fortification from the front, an enemy could only discern the slope of the glacis and the slope atop the parapet. If an enemy were to attack in the open against such fortification, he would come under precisely sited cannon and musket fire. All glaces were covered by fire from a curtain wall and from the flank by supporting bastions, producing a crossfire. The gentle slope on the top of the higher parapet made the musketeers on the firing platform discharge their weapons at a precise angle calculated to have the musket ball follow a specific trajectory where it would fall at a precise point against advancing enemy on the glacis. Officers instructed each soldier to discharge their weapon only at a perfect perpendicular angle to the line of the wall upon which they stood in order to ensure that the musket balls of a line of firers standing side by side were evenly spaced when they fell upon the glacis. So long as the defenders kept good discipline in the discharge of their muskets, there were no places safe from fire for the attacker.[50]

Vaubanian constructions used ramparts, ditches and a glacis, bastions, and all of the features perfected during the two previous centuries—barbettes, embrasures, covered ways, war pits or *trou de loup*, palisades, abbatis, gates, barriers, redans, lunettes, redoubts, and countermines. Cadets read about each of these features in the *Treatise*, recited their characteristics, and displayed understanding of these fortification features on the recitation room chalkboards and in their notebooks for grading by the instructor.[51]

Field fortifications differed from permanent works by size and by material used, but the basic principles remained the same. Differences in size and material also meant a difference in the complexity of mathematics needed to design and calculate projections, materials, and labor requirements. The entire First Class learned such calculations for field fortifications, which were made from earth and wood. Design and construction plans for stone and masonry fortifications were much more difficult and taught to the First Section only. Vauban had originally made this distinction.[52] After just a few months of training, siege engineers could dig trenches and construct field fortifications. Permanent fortifications engineers, however, required rigorous education and years of experience to learn how to overcome variances in local topography, hydrology, and mineralogy and to determine the best methods for stone cutting and construction suitable to each locale. Such skill was essential for Americans destined to build coastal fortifications along an extended and highly varied coastline with diversified weather conditions. Instructors at West Point expected cadets of the First Section to demonstrate how to apply to fortifications building the basic knowledge of hydrodynamics, chemistry, mineralogy, and topographical sketching learned in the third year of study.

Cadets of all sections were expected to understand how to site artillery within fortifications and counter siege artillery fire. Cadets of the First Class needed to demonstrate complete knowledge of the artillery lessons taught throughout their third year. They were also taught siege techniques, which had changed very little since Vauban had perfected his "method" in the late seventeenth century. This involved several distinct phases. The first phase was to completely surround the fortification to be captured and encamp the besieging army within its own circular fortification outside of cannon range of the fortress. Once this encircling contravallation was complete, the next phase commenced with digging zigzag trenches from multiple points from the contravallation walls toward points of the fortress wall that the attackers eventually wished to breach. This systematic use of advancing zigzag trenches avoided the bloodbath that would happen in any open attack upon such defenses. These zigzag trenches allowed for the safe forward movement of soldiers and cannon toward a particular point of attack. At various ranges from the walls, attackers built trenches parallel to the fortification. Each parallel served as a place to array large numbers of troops and artillery batteries (a *grande place des armes*), to store supplies, and to provide suppressing fire onto the walls while the zigzag trenches advanced. Just as the use of bastions concentrated the fire of the defenders on the flanks of an attacker, Vauban's parallels applied the same principle of concentration of fire. Mutually supporting batteries, firing from enfilade or "enveloping" positions on the parallels, produced a concentrated effect on the point in the fortification where a breach was to be made.[53]

The final and decisive phase of the siege happened when the attackers captured a portion of the covered way and ditch. This allowed for the positioning of artillery on the covered way at point-blank range to the exposed curtain wall, which was open to fierce battering and susceptible to collapse, forming a breach. Vauban calculated that most fortresses would fall to his method within forty to fifty days.

Vauban's siege technique produced such certainty of success that he had called it the "triumph of the method."[54] He maintained that with it "Dame Fortune" played far less a role in the attack of fortifications than did "prudence and dexterity" in the application of the method.⁵⁵ Vauban had sought certainty of success to reduce unnecessary bloodshed, something reiterated by Professor Mahan in his teaching at West Point. Both men vigorously applied mental reasoning to warfare in an effort to make it economical. Neither held reliance on individual courage as a substitute for careful thought. In the 1820s Prussian theorist Carl von Clausewitz agreed, stating that "siege warfare gave the first glimpse of the conduct of operations, of intellectual effort."⁵⁶

This idea that warfare required intellectual effort provided O'Connor with a framework for his appendix, a summary of the principles of *grand tactics and operations*. Here O'Connor compiled some of the best ideas of les Lumières but emphasized those of Jomini. He introduced the appendix by presenting Jomini's idea that commanders use geometry to discern the geographic theater of war by visualizing the various frontiers of importance as a relationship of lines (parallel, perpendicular, oblique) that would become the bases of operations for the opposing armies. To "place this idea in clearer light," Jomini asked that the reader consider the theater as a four-sided figure, a square, a parallelogram, or a trapezoid, with opposing frontiers making up two or more of the sides. He acknowledged that if one army possessed several of the sides of the theater of war and could employ them as bases of operation, that army would gain an advantage. In this prescription, Jomini used (but never admitted to it) a quadrilateral to substitute for Bülow's triangle—with the same effect of increasing room for advancing army to maneuver.

A frontier became a proper base of operations when secured by one or more fortifications that served as depots (*grande place des armes*) from which an army might march into the enemy's terrain or retreat upon. The base served the same purpose as a line of contravallation in a siege. It was a secure area where an army could chose points to stockpile supplies, ammunition, and weaponry. The advance from the base (not unlike the zigzag siege trench) was to be upon a line or lines of operation to concentrate forces at an objective point, the possession of which would disadvantage the enemy's army; not unlike the disadvantage realized by defenders when a besieging army captured a portion of the covered way allowing for concentrated fire on the exposed curtain wall.

An attacking general would need to acquire information about an enemy's frontier in order to select the best lines. This was decisive, for a bad choice of this line would be fatal. Three principles therefore governed the appropriate selection of such a line: keep it short to achieve the fastest "velocity" of movement; choose a line that an enemy cannot flank or intercept; and choose a line that leads to "some decisive object," preferably one that intersects an opponent's line of communication. A commander could establish intermediate supply bases or depots on the line of operations that would also serve the same function as a siege parallel, offering a place to stock provisions that would grant velocity to advancing troops, that would secure lines of communications, and that could serve as winter quarters for an invading army.

In emphasizing geographical lines upon which armies operate, the points of importance for supply, security, and quartering, and an ultimate geographic objective point, O'Connor defined the elements of campaign design that would last until the advent of the aircraft. These principles relied on knowledge of topography and geometry and resembled Vauban's siege methodology. O'Connor's theory recommended angular projections from secure lines of defense using incremental movement toward an objective and the steady concentration of military forces on a single point. The focus of Vauban's concentrated effort was the point of breach in an enemy's walls. The focus of effort in nineteenth-century campaign theory was a point along an enemy's line of communication that would sever him from his base and force battle at a disadvantage. As such, turning movements (maneuvers that turned an enemy's flank) were essential. Napoleon had endorsed this reasoning and summed it up: "The principles of war are the same as those of a siege. Fire must be con-

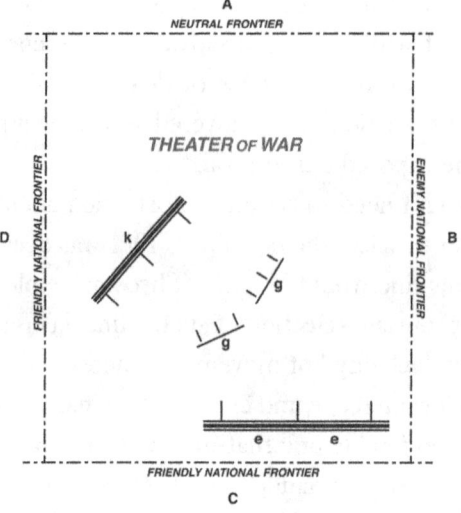

FIGURE 4. O'Connor's reproduction of Jomini's "Configuration of a Theatre of War." Based on a sketch by Capt. John O'Connor from *Treatise on the Science of War and Fortifications Composed for the use of the Imperial Polytechnick School, and Military Schools, and Translated by the War Department for use of the Military Academy of the United States to which is added a Summary of the Principles and Maxims of Grand Tactics and Operations* by Simon François Gay De Vernon, translated by Capt. John O'Connor (New York: J. Seymour, 1817), 434.

centrated on a single point and as soon as the breach is made the equilibrium is broken and the rest is nothing."[57]

A *Plan of Campaign* required choosing the best line of operations to bring about a concentration of mass on a point of vulnerability or, if on the defensive, to deny the enemy's ability to concentrate on vulnerable points. The idea of the single point of concentration pervaded O'Connor's summary:

> The secret of victory is to operate with superior numbers or masses against one or two points, or to bring the whole or greater part of your force to bear upon a point where the enemy has only a small portion of his to resist you; and where, if victorious, you will take in flank or reverse his whole line or position.[58]

To achieve concentration, a general should attempt to mask marches upon a flank and move with "celerity and in mass." Only celerity could put an army in a position of advantage over an adversary.[59] Cadets learned the importance of "stealing marches," choosing lines of operation that allowed faster movement than an enemy, thereby seizing and maintaining the initiative. This "science of manoeuvres . . . [was] called *manoeuvring against time*; for the army that manoeuvres with the greatest celerity, must, all things else equal, vanquish its adversary."[60]

> The science of war is founded upon *concentration of force* and *celerity of movement*. Consequently the great art is to put the greatest mass of troops in simultaneous action against such a point in the enemy's line of battle or operations as threatens his flank and rear; and where, if successful, he can hardly escape capitulation as destruction, and being cut off from his communications . . . assume the initiative on the offensive, and not to give the enemy time to combine any movements against us.[61]

Emphasis on maneuver did not merely entail marching but put a premium on road and bridge building and on skills in organizing and transporting armies and supplies.

The importance of seizing the initiative also prevailed in contemporary thinking about defensive operations. The best form of defense was in fortified frontiers that allowed the conduct of limited offensive maneuvers that spoiled an enemy's offensive capability. O'Connor referred to this in the contemporary term *active-defense*. It would become a key principle in Confederate war strategy.

O'Connor's summary gave a brief survey of the evolution of the theory of the base and lines of operation established by Lloyd and Bülow before presenting Jomini's descriptions of the multiple types of lines of operations—single, double, multiple, interior, exterior, long or deep or remote, concentric, eccentric, secondary, and accidental.[62] O'Connor favored Jomini's opinion of lines, particularly his preference for singular lines of operation operating from a frontier with massed forces interior to an enemy's more dispersed army. For O'Connor, Jomini had "transcended all writers in war. . . . He has reduced the hitherto mysterious science of war to a few self-evident principles and axioms."[63]

O'Connor condemned the use of exterior lines against an enemy who had a central position, and he stated that even with the use of two frontiers, an army should concentrate in a central position in the theater of war.[64] He dismissed alternatives allowed by Bülow, by Archduke Charles, by Napoleon, and by Jomini himself. In his interpretation, he rejected the idea of using a long base of operations with multiple lines of operation emanating from the extremities toward a common object point—thus creating wide angles of operation. O'Connor instead held Frederick the Great's Rossbach campaign as the epitome of concentrating mass in a central position between dispersed enemies and the defeat of that enemy incrementally.[65] His summary, as with most texts of this period, was full of historical examples, and he reproduced Jomini's depiction of Napoleon's 1806 campaign to reinforce his preference for operating along a single line of operations to achieve a central position.

O'Connor's emphasis on Jomini's lines of operation theory was more suitable for European officers than for West Point cadets. The vast size of the United States—with its long and diverse inland and maritime frontiers—made Jomini's prescriptions of the central position on interior lines far less relevant than in continental Europe. With such vastness, maneuver options were plentiful, making reasoning about preferred lines of operation nonsensical. American commanders during this period already knew this. Winfield Scott, Edmund Gaines, and Andrew Jackson were less inclined to follow the pedantry regarding the preference of interior lines, and they demonstrated this in their own campaigns. America's chief military theorist of the antebellum era, Professor Mahan, would also be more open-minded about the variety of lines of operation possible in the American context.

Everyone agreed, however, that bases of operation (or bases of supply) were essential and that topography dictated where good bases and lines of operation could be situated to allow for concentration of force. In the various campaigns fought by the U.S. Army in the antebellum period against Native Indians and the Mexican Army, bases and lines were employed, but the use of a singular interior line of operation to gain Jomini's coveted central position was never a key factor in plans. Few U.S. Army officers, with the later exception of Henry Halleck (see chapter 7), were ever as limited in their thinking as was O'Connor. That said, few grasped how difficult it would be in vast North American theaters of war to force a decisive destruction of an enemy army. That knowledge only came in the Civil War. Before then, West Point instruction under Professor Mahan would provide alternatives to O'Connor.

In terms of battlefield tactics, Gay de Vernon and O'Connor simply followed Napoleonic prescription. The artillery opened fire "within good distance," and light troops followed up with their own fire while advancing forward. Line infantry delivered maximum range fire and then formed into columns to attack.[66] The optimum form of attack was on an enemy's flank, and Gay de Vernon described variances based on angles of the opposing lines—parallel, the slight acute, the acute (enfiladed or outflanked angle), the right angle (described as "taken in the flank"), and the obtuse angle ("taken in rear or cut off").[67] The linear nature of tactical thinking was necessary; lines provided maximum effective use of individual musket fire. Regardless of angles of advance, Gay de Vernon advocated that troops who broke under attack receive no quarter but be pursued and destroyed in detail. Audacity and fighting spirit mark this text.[68]

In his translation of Gay de Vernon, O'Connor tried to discern for the cadet a complete military science, knowable to anyone who was prepared to study. He did not attempt to hide simplifications and reductions: "The idea of reducing the system of war to one primitive combination, upon which all others depend; and which should be the basis of a simple and accurate theory, presents innumerable advantages . . . it should regulate all plans and actions."[69] This idea was presented not to provide certainty of outcome, nor to make war follow a particular rule or matrix, but to suggest valid and effective combinations in order to render the vast complexity of war intelligible, to reduce the mystery of war by applied mental faculty, and to reduce reliance on costly experience or genius:

> [Anyone] who would attribute everything to natural genius, or to accident, may perhaps cite several that are exceptions, and which succeeded by contrary principles. But they are mistaken.... Genius is undoubtedly a great share in victory, because it presides over the application of acknowledged rules, and seizes all the modifications of which this application is susceptible. But in no case will a man of genius act in violation of these rules; and *he* will never be acknowledged as a great captain, who has won a battle by accident and against the rules of the science; ... such a victory is only a proof of reciprocal incapacity; of a total absence of tactics.[70]

In their inexperience, cadets would probably not have appreciated the depth of meaning in O'Connor's treatment of genius, but they recited his work nonetheless.

Four years at West Point exposed cadets to a "complete system" of warfare. Training in tactics and summer encampments taught duties of the soldier and junior officer, as well as company, battery, and battalion drills, and how to live on campaigns. The academic curriculum provided knowledge about where to find answers to technical questions regarding the manufacturing of arms and munitions, the construction of wagons and gun carriages, the building of bridges and roads, the encampment and marches of armies, and the importance of supply depots—all essential to Calhoun's idea of the officer corps needed in an "expansible army." Cadets learned to apply basic principles when siting fortifications, reconnoitering frontiers, moving armies, and arranging infantry, artillery, and cavalry for battle.

The academy instructors expected cadets to embrace this system of thought. An Enlightenment institution, the U.S. Military Academy came as close as an institution can come to encouraging the use of Pascal's *mathematical mind* to solving military problems in varying circumstances. The system may appear stifling to twenty-first-century readers, so accustomed are we to dialectic. Some historians have also suggested that the system was anachronistic. Yet Gay de Vernon and O'Connor were, if anything, entirely Napoleonic, which is to say their system was state of the art. They did not harp back to wars of the eighteenth century, nor did they prescribe avoidance of battle or slow deliberate and mechanistic maneuver. In Gay de Vernon and O'Connor, celerity of movement and concentration of forces to destroy an enemy remained sacrosanct. However, they were also wise enough to admit preference for the severing of an enemy's line of communication to his base of operations as a more assured means of destruction.

Cadets left the U.S. Military Academy with sound knowledge of fortifications and military engineering, artillery tactics and ordnance functions, and battalion logistics. Cadets also understood the theory of how to move an army on a campaign of *la grande guerre* and especially the importance of strategic lines and points. The curriculum was indeed well rounded. With the exception of lessons in strategy, the focus of instruction was on organization and tactics and the practices necessary for mobilization and preparation of the nation for war. The mandate of the academy was to produce a lieutenant, but the expectation was that the lieutenant would be capable of employment in a variety of line and staff positions. This was important because in the 1820s a practice began of seconding a considerable portion of the line officers to the staff. This portion grew to an average of one-quarter of all officers between 1830 and 1860 as the size of the army, and its tasks, increased.[71]

4

Internal Improvements

Military science had barely become doctrine in Sylvanus Thayer's academy when faculty began to modify it to meet the changing requirements of the United States in the 1820s and 1830s. The content of the 1820 curriculum pertained to *la grande guerre*, emphasizing fortifications and artillery applications that might counter an amphibious threat from a European power. The Atlantic perspective that dominated ideas of national defense began to change in the mid-1820s when peace with Europe allowed Americans to focus their energies inward. Public desire to improve transportation infrastructure and westward expansion created competing demands for engineering and survey skills. The army met these demands by increasing the number of officers employed in civil engineering and survey duties and by increasing its staff capacity to deal with the rising number of projects and military posts established in the West. The War Department created a Board of Internal Improvements to oversee the new activity, whose members were the same as the Board of Fortifications. The coordination between these boards and the staff bureaus created a synergy in strategic thought about integrated national defense. The department and the boards also supported a shift in emphasis of West Point curriculum and the academy's military science expanded to include civil engineering, surveying, and reconnaissance duties.

The Shift to Internal Improvements

Europeans in the 1820s reaped the benefits of the Congress of Vienna and for the first time in a generation pursued peace. A beneficiary of this peace was the United States, where the reduced threat of invasion or interruption in trade allowed people to concentrate on economic growth. Immigration, expanding cities, increased agricultural production, and the beginning of the Industrial Revolution led to demands for improved transportation systems. Responsibility for interstate roads and waterways could not reside with the states themselves, and the only federal agents authorized to work within state boundaries were postal and customs personnel and those army officers conducting construction of the Third System. Such officers were already working with state and private business leaders on fortifications tasks, so it was logical for the government to ask the War Department to also assume responsibility for planning projects that contributed to a national system of transportation—ostensibly for purposes of defense, but in actuality to the economic, social, and political benefit of all. The government referred to these efforts as "internal improvements."

Internal improvements initiatives were not new. Washington, in his 1783 "Sentiments," had stated the requirement for "military roads" to move mustered militia forces to threatened points on the frontiers. Gen. Anthony Wayne had cut roads into Ohio in his 1794 campaign.[1] President Jefferson had perceived it as a federal responsibility to improve the related functions of "public education, roads, rivers, canals, and such other objects of public improvement."[2] Secretary of the Treasury Albert Gallatin captured this idea in part in his 1808 "Report on Roads and Canals."[3] After the War of 1812, successive secretaries of war sought to construct military roads under the auspices of postal delivery, a federal responsibility. The first "national road" was constructed incrementally, between 1811 and 1818, extending from Cumberland, Maryland, to Wheeling, Virginia. However, a question of constitutional authority hampered all these initiatives, and Republican presidents remained reluctant to force federal projects within state boundaries.[4]

The War of 1812 had demonstrated that, especially on the northern frontier, the lack of military roads was disastrous. There was simply no link between settlements where troops and supplies could muster and the military frontier where they should engage the British. Without adequate road and water transportation networks, there could be no viable base or lines of operations for military actions in the Northwest and few on any other frontier. For this reason, Secretary Crawford suggested in 1816 that the regular army build roads in times of peace, leading Generals Jackson and Macomb to begin road construction without legislative authority, using general military funds.[5]

In 1818, Secretary of War Calhoun revamped federal transportation initiatives. His 1819 "Reports on Roads and Canals" emphasized that a system of internal communications was a necessary corollary to coastal defense and drew attention to the relationship between systems of national defense, inland transportation, and public commerce.[6] The 1821 "Report on Fortifications," which laid down the first integrated defense policy, highlighted "interior communication by land and sea" as the third of four critical components to the national system (alongside a protective navy, frontier fortifications, and a regular army and organized militia).[7] Interior communication included road building, harbor improvements, river dredging, bridge building, and canal construction. It would later include the building of railways.

The federal government discussed internal improvements at a time of public frenzy for canal building. As one engineer reported, "Of all the means which human ingenuity has devised for facilitating communications . . . canals occupy, at present day, the highest rank."[8] Americans wanted to transport people and goods across the great stretches of land and wilderness of the expanding nation. At that time the Erie Canal was nearing completion, linking Lake Erie with the Atlantic seaboard, and the corps of engineers was recommending plans for canals to connect the Illinois River with Lake Michigan and the Des Plaines River and from there to the Mississippi. The advantages of such works for both commerce and defense were self-evident.

Federal focus on internal improvements also promised to facilitate the steady movement of people toward and across the Mississippi. However, before roads and canals could be built to assist such movements, there was a requirement for reconnaissance and survey work in new territories, and this task also fell to the regular army. President Jefferson had set a precedent for this by sending Captains Meriwether Lewis, Richard Sparks, and Zebulon Pike on expeditions to chart and map in advance of expanding American frontiers. In 1818 the War Department sought to regulate westward exploration by establishing a topographical bureau of ten engineers (a separate component under the Engineer Department), with a mandate to collate the work of surveyors, explorers, and cartographers.[9] In the race with the British to secure lands beyond the Mississippi, the War Department sponsored a series of western expeditions. Most of these (perhaps best represented by Maj. Stephen H. Long's 1817 and 1819 exploration) included a scientific component that used natural philosophy, sketching, and statistics to catalog the geology, flora, fauna, and inhabitants of the central plains and western mountains, including estimates of warriors living in these areas, with recommendations for how best to open up such territories.[10]

The role of the topographical engineers (commonly referred to during the antebellum period as "Topogs") was not simply to serve the civilian desire for surveys and exploration. The original order establishing the bureau reflected its role in *la grande guerre*, directing its members to "make plans of all military positions . . . to accompany all reconnaissance parties sent out to obtain intelligence" and to "exhibit the positions of contending armies on fields of battle."[11] However, in the 1820s the secretary of war and the senior engineers in both the Engineer Department and the topographical bureau saw that the military purposes of the Topogs and public demand were not mutually exclusive. If the survey work of the Topogs aided territory, state, and federal plans for improved transportation and westward migration, they also fulfilled the military requirement for surveying and planning military frontiers and lines of communications. While conducting survey work for internal improvements and territorial expansion, the bureau never lost sight of its military purpose and role as "an essential branch of the general staff." A 1827 bureau report couched this purpose: "It is by means of their reconnoiterings, maps, and descriptive memoirs, that the importance of military properties of positions are known, and that the strategic circumstances of a country are ascertained." The report then clarified the role of the Topogs: "Modern armies cannot move without a perfect knowledge of the geography and topography of the field of operation, and it belongs to the topographical engineer to furnish all information calculated to assist the general in the research of local circumstances, in order that he may be able to plan his operations accordingly."[12] Topog work was therefore important for military movements into the West.

To assist in fulfilling this general staff function, the topographical bureau became the repository of all reports and memoirs, maps and charts, "itinerary tables respecting the concentration of militia at points of rendezvous, and military histories of past operations to help the government in planning military combinations that other wars might call for."[13] The War Department had begun as early as 1821 to amalgamate surveys conducted by the Fortifications Board with federal and state surveys conducted by the topographical engineers, keeping everything within the topographical bureau in a manner similar to the way that French Minister Lazare Carnot had done in his Bureau Topographique in Revolutionary France.[14] While Topog surveys and reports might support the development of civil infrastructure and the progress of science, their worth to military planning was recognized.

The Third System, internal improvements, and western expansion created constant demand for engineering skill, exacerbated in 1822 when the topographical bureau's chief, Maj. Isaac Roberdeau, suggested that the federal government assume responsibility for coordinating all matters of civil works in America, from education to employment and regulation.[15] The entire engineering capacity of the War Department—twenty-two military and ten topographical engineers—was already stretched in running the military academy, in building coastal fortifications, in improving coastal harbors and inland waterways, in surveying the west, and in cataloging the nation's natural resources. When the president approved the General Survey Act of 1824, demand for engineering skill increased further. The act allowed the president, without consulting the states, to employ military officers and civil engineers to "cause the necessary surveys, plans, and estimates, to be made of the routes of such Roads and Canals as he may deem of national importance, in a commercial or military point of view, or necessary for the transportation of the public mail."[16] The act did not authorize the federal government to build infrastructure, only to survey and plan. Nonetheless, it gave the government the mechanism to coordinate construction projects between the states and to integrate further the system of national defense.

Under such presidential authority, the Engineer Department established the Board of Engineers for Internal Improvements in May 1824, with Brig. Gen. Simon Bernard, Col. Joseph Totten, and civil engineer John L. Sullivan as the standing members. They assigned twenty-four military officers and civil engineers to the board's employ. However, even this was not enough to meet the demand.[17] By the end of 1824, the board employed twenty-six officers. In 1825, forty were employed, and in 1826, fifty-three officers were under orders from the board. Approximately fifty officers would remain dedicated to the work of internal improvements for another decade.[18] At any given time, the board controlled a dozen internal improvement projects, separate from those of Third System fortifications.[19] The demands on the Engineer Department were severe, and in January 1826 Congress agreed to increase the number of junior officers of the engineer corps and to triple the size of the topographical bureau.[20] It would take another twelve years to find the funds to make this actually occur.

The demand for civil engineering skill rose further after the opening of the Baltimore and Ohio Railroad in 1828, when railroad construction started to usurp canal building as the nation's biggest infrastructure effort. The federal government desired to regulate railway development by assigning regular army engineers to oversee projects, an effort that continued for the next forty years.[21] In this, the department competed with state and private enterprise, which shamelessly recruited army engineers and West Point instructors for civil employment. The lucrative nature of civil engineering and railroad work seduced many officers into state and private service. Professor Crozet left West Point in April 1823 to become a principal engineer and surveyor of public works in Virginia.[22] The next head of the Engineer Department, Capt. David Douglas, resigned his commission in 1831 to assume engineering work with the Morris Canal Company of New Jersey. Douglas could hardly afford to ignore the demand for his expertise. His salary as a captain of engineers was $1,239 a year, while as a civil engineer he could easily make $4,000 for three months' work.[23]

The secretary of war looked to the military academy to provide the proposed increase in officers to meet the new demands: "It is from the Military Academy alone that we must look for those who may possess the requisite acquisitions and talent for the purposes of either the military, topographical, or civil engineers."[24] West Point was not merely an obvious source. It was the only choice. Civil engineering in the United States, as in England, had not reached the status of a distinct and distinguished profession as it had in France, with her multiple civilian engineering schools and associations.[25] Until 1824 West Point was the only engineering school in the United States. That year Stephen van Rensselaer and Amos Eaton opened the Rensselaer School (but Eaton sent his own son

Amos to West Point in 1826). Not until 1840 did Rensselaer or any other college begin to challenge West Point in numbers of engineering graduates. Harvard and Yale did not open their engineering schools until the mid-1840s, but engineering education did not become relatively common until after the Civil War.[26] This fact perforates Samuel Huntington's assertion that antebellum West Point's civil engineering focus replicated that of civilian colleges, robbing West Point graduates of anything to distinguish them from civilians.[27] Until the 1860s, the West Point curriculum was unique, and graduates were coveted by society.

Federal pressure on the military academy in the 1820s to produce more engineers met with opposition from the academy staff. The faculty (backed by the chief of engineers) refused to lower its standards to produce more engineers without increasing its cadet body. Despite proposals to expand the corps, the congressional Committee on Military Affairs was unable to find ways and means to increase the number of either military or topographical engineers or to create higher academy output.[28]

The faculty found an expedient by shifting emphasis within the curriculum to civil engineering, ensuring that all graduates had some capability in this field. A second expedient was the employment of more and more infantry and artillery graduates in surveying. By 1826 thirty-two line officers were assigned to work alongside the twenty engineers and Topog officers for the Internal Improvements Board.[29] This practice continued until 1838, when an expanded corps of Topog engineers took over such duties. Between 1824 and 1837, Topogs completed 107 works of internal improvement authorized by Congress, disbursing $14,430,614. They also completed 263 surveys and examinations and published 776 drawings.

Military historians have failed to see the connection between the work of the army in meeting public demand for surveying and civil engineering and the policy for national defense, a connection that explains why the academy expanded its military science to include fields of study needed for the army's work in internal improvements. Some, following Huntington's thesis, see civil engineering as something that diverted the army's energy away from the higher need to prepare for war.[30] However, these opinions ignore the consistent and evident sentiment in the War Department throughout the 1820s and 1830s. In an era when states' rights and sectionalism were emerging as powerful forces to limit federal activities, and when civil demand for improvements steadily grew, the War Department continued to see the nation as a single whole, requiring an integrated defense policy that precluded the need for a large standing army. The department refused to dilute power by allocating tasks to states' militias. Instead, they continued to use the army as an agent to promote integration within a defense policy that saw the nation as contained within military frontiers, requiring improved networks of communication internal to these frontiers for both military and civil purposes, all part of a single strategic problem requiring an engineering solution.

Besides, all officers involved in such meaningful activity developed professionally. Internal improvements forced officers to represent federal interests when working beside state and business leaders to secure procurement, transportation, and construction contracts, providing them with wonderful experience that would serve them well when generating and supplying state volunteer armies for war. When graduates performed internal improvements duties, they acquired higher skill and knowledge of reconnaissance work and logistics. These were precisely the skills that in times of war would allow them to function more easily as commanders and general staff officers. Of the 1,966 graduates of the academy between 1802 and June 1861, 326 were employed under the Fortifications Board or under the Internal Improvements Board performing Third System or internal improvements engineering duties.[31] Of these 326, 189 were infantry, artillery, or cavalry officers borrowed from line units. The heaviest period of demand for seconded officers was between 1818 and 1838. One-quarter (180) of the 708 academy graduates of this period were employed in such work. The personal and professional benefits of this service have not been recognized in the historiography. This work allowed officers the opportunity to apply the military science they had learned at West Point and to hone professional skills. Senior Topog officer Col. John Abert later stated that any officer involved in such tasks was "perfecting himself in the practice of his profession acquiring exact and persevering habits of investigation, improving his *coup d'œil*, and gathering the most valuable information in relation to the capabilities of self defence of a locality . . . and its ability to aid in the defence of other parts of the country."[32]

Simon Bernard and Joseph Totten ran both key engineer boards, giving the War Department a means to coordinate military and civil engineering efforts for national defense. The close connection was first evident in 1824 when the Fortifications Board provided the Internal Improvements Board with its first list of twenty-seven possible canal projects, rating each for both its military and its commercial value.[33] Their work thereafter was carefully coordinated, in close cooperation with the quartermaster general, who oversaw construction of inland garrisons, depots, and military roads and bridges.

The demands of the Third System and of internal improvements put strain on all other staff bureaus. To ensure adequate staff capacity, each of the bureaus began in 1821 to receive officers detached from line units to serve in staff positions for one or more years. West Point graduates were preferred. Over 25 percent (504) of all graduates of the academy between 1802 and 1861 served for some time under the quartermaster general, the adjutant general, the chief commissary, or the chief ordnance officer.[34] Just as with those officers serving under the engineer boards, these officers routinely applied knowledge of military science learned at West Point and benefited from their staff experience by being better prepared to raise, equip, move, and supply large volunteer armies.

Historians have paid too little attention to the work of the staff bureaus, particularly that of the quartermaster general. The U.S. Army in the antebellum era was distinguishable for the number of officers dedicated to quartermaster work and the army's emphasis on proper logistical planning. In this, the army surpassed West Point instruction, producing every few years more detailed quartermaster regulations that officers implemented each day in garrisons and camps throughout the United States. The 1841 edition of the regulations captured the scope of quartermaster work thus:

> The objects of this department are to ensure an efficient system of supply, and to give facility and effects to the movements and operations of the army . . . provide quarters, hospitals, and transportation for the army, . . . camp and garrison equipage, . . . direct the survey, and superintend the opening and repairing of roads, and the constructing and repairing of bridges . . . necessary to the movement of any part of the army . . . to provide storehouses . . . [all] purchases . . . to provide material and direct and superintend the constructing and repairing of quarters, barracks, hospitals, store-houses, stables, and other necessary and authorized buildings . . . blockhouses and other necessary defenses . . . select sites of encampment . . . direct the movement, and be responsible for . . . supplies required for . . . the army in the field.[35]

The integration of the work of the engineer boards and the staff bureaus made strategic planning easier. War Department officers worked together to select strategic points and lines of communications, both along the frontier and in the interior. Engineers sited proper locations for fortifications on the coastal and inland frontiers, while officers serving as Topogs sited and surveyed roads, bridges, and waterways that linked fortifications to national depots and points where volunteer armies mustered. Officers acting as post, district, or department assistant quartermasters oversaw the construction of the military infrastructure using civilian contractors. Meanwhile, ordnance and artillery officers and officers acting as assistant commissariat and assistant adjutants general developed plans to equip, arm, move, feed, and supply regular and volunteer soldiers deploying along the frontier or on the lines of communications.

The coordinated work of the boards and staff bureaus granted strategic character to the implementation of the War Department's policies, involving mutually supportive military, political, and commercial interests. The integrated board structure was also a perfect system to sustain public support, because it served both state desires for improvements and public demand for surveys and roads without threatening the fragile balance between states' rights and federal government obligations. Yet the work of the boards at the same time remained directly related to preparation for war. Because of this, army officers began—and continued throughout the antebellum era—to think about the problems of fortifications, transportation, and logistics on a continental scale. This strategic perspective forced many officers to develop professionally well in advance of state volunteer officers and even in advance of European officers of equivalent rank.

In boards and bureau planning, the roles of the regular army and the state militias remained distinct parts of a system that made perfect sense and avoided the traditional fears associated with a large standing army. The U.S. Army of the 1820s was perceived not as an institution that was "consuming the labor of others" like their European counterparts, but as "a body of military and civil engineers, artificers, and laborers, who probably contribute more than any other equal number of citizens," dedicated to "the progressive execution of our great system of national defence."[36] Despite Samuel Huntington's portrayal of the antebellum officer as living on the margin of society, evidence suggests that public respect for army officers grew throughout the antebellum period.[37]

Frontier Military Roads

While not officially part of the Third System or of Internal Improvements, the building of frontier "military roads" also took on a strategic character over time. Starting in the 1820s, once they began to push beyond the Mississippi valley, settlers looked to the army to build and police roads and posts that would better protect them from Indians and, as an added benefit, provide economic opportunities. This activity fell outside of the responsibilities of the engineering boards, but came under the authority of the quartermaster general. Line officers on detachment to his bureau built roads and sited and constructed frontier forts and garrison towns. These efforts were supposed to be coordinated within the overall strategic system of defense, but this was often difficult because of public demand for local roads that had only commercial value.

In 1824 the War Department authorized road building from the Mississippi River to Little Rock, Arkansas, and from Pensacola to St. Augustine, Florida. In 1826 Michigan raised demands for military roads and posts to counter British-Indian threats between Detroit and Chicago. Military roads were built in Maine in 1829, and Arkansas extended its networks in the early 1830s. Demand for more roads remained constant. A memorial from citizens of Clay County, Missouri, for instance, presented to the Senate and House in December 1835, suggested that one road connect "a line of military posts ... established along or near the boundary between the settlements and the Indians, beginning on the upper Mississippi and extending to Red River: that well constructed and permanent forts be built sufficiently large to garrison one company of infantry and two of dragoons."[38] The War Department responded by commissioning surveys for what would eventually become known as the "Great Military Road" project, which envisioned a road running from Fort Jesup, Louisiana, to Fort Snelling, Minnesota.

Army officers debated the plans for this road. The War Department initially supported civilian suggestions for a south-north orientation, joining frontier forts along a clear path patrolled by dragoons. However, Acting Quartermaster General T. Cross deemed such an orientation to "violate a fundamental principle of military science." Roads needed to connect frontier garrisons with depots in the interior and not run along the frontier itself: "the lines of communication should be *perpendicular* to the frontier, not *parallel* with it."[39] The road network should not run south-north but be more like the

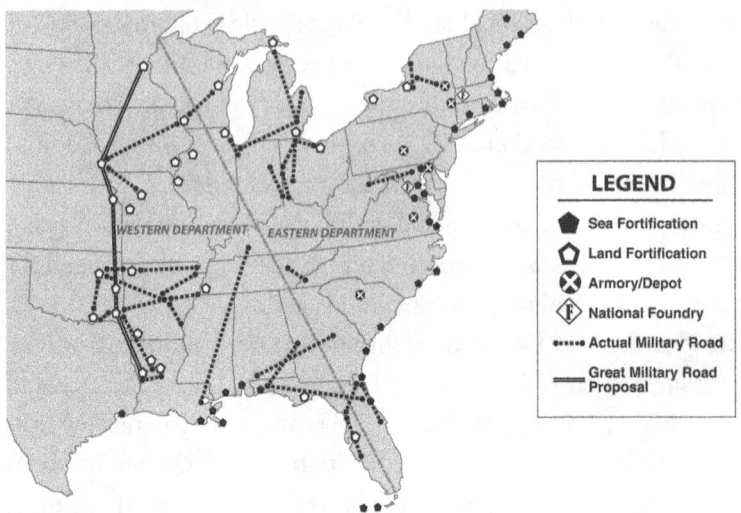

MAP 4. Major military roads built, 1818 to 1840. Information compiled from various reports in ASP MA, including "A Report for the Fortification of the Seacoasts" by Brig. Gen. Simon Bernard, Maj. Joseph Totten, and Capt. J. D. Elliot (U.S. Navy) 2:304–13. The concept survived until the Civil War and was reiterated and modified in detail periodically (see ASP MA, 3:327 (February 25, 1826), 283–302. Chief Engineer Alexander Macomb delivered a complete plan, including revised estimates for each fortification site and garrison requirements for both peace and "siege" conditions.

spokes of a wheel with westward projections running from central interior points. Whichever form the road took, the department estimated that with 66,499 Indians along this frontier, 5,000 soldiers would be required to secure it. The debate over orientation resulted in the road being only partially completed. What is interesting in this matter is how much the debate exposed the breadth of strategic thinking in the War Department regarding matters of regional defense. Road construction was not a simple affair of surveying and construction to suit local conditions. Bureaus attempted to assess the strategic value of each project, made estimates of resource and garrison requirements based on empirical threats, and attempted to consider the value of each project to the whole system.[40]

Despite historians' generalizations that western frontier duty was always drudgery marked by the occasional exhilarating skirmish and that nothing at West Point had prepared these officers for this work, military science had its place on the western frontier and kept a portion of the officer corps busy. Infantry and artillery officers serving as acting assistant quartermasters in each post and regiment (under the quartermaster general's authority) were responsible for planning and constructing roads and posts.[41] Officers charged with the construction of these forts used their West Point engineering courses to full measure. They even applied their civil engineering knowledge to the construction of family domiciles. The civil and military engineering requirements of service on the frontier could only be met by men educated and trained for this work. The West Point academy ensured that such skills remained resident wherever units of the line encamped.[42]

West Point Civil Engineering in Context

As internal improvements and military road projects increased, expectations about the role of the military academy expanded. As early as 1821, members of the Board of Visitors voiced the opinion that "this academy may, without expense, be made to furnish Civil engineers for roads, bridges, canals, and hydraulic constructions, as well as Military and Topographical Engineers." The board understood the relationship between the unique academy curriculum and internal improvements: "The Military Academy at West Point is the only establishment where a complete system of Instruction in Mathematicks, Descriptive Geometry, Mechanicks, Astronomy, and Civil and Military Architecture is procured and which afterwards places the pupils in situations, where they have an opportunity of exercising the different Civil and Military avocations, to which the application of exact sciences is indispensible."[43]

The 1830 Board of Visitors added that a key purpose of the academy was to "furnish science for exploring the hidden treasures of our mountains."[44] Clearly not all observers saw the military purposes of the academy as predominant. While tactics remained the top priority, civil engineering subjects gradually received more emphasis in the late 1820s. The War Department understood and approved of this. The secretary of war stated it this way in 1825: "To the military science proper are added many auxiliary acquirements . . . particularly civil engineering, which, from the efforts everywhere making by the general and State governments for internal improvements, promises to be amongst the most beneficial acquisitions in the whole range of science."[45] The direct effect of this focus was a gradual expansion of civil engineering topics taught to the First Class in their first term, complemented during that term with a course in drawing, revised in 1825 to include sketching and drawing and shading "elements of landscape" and "elements of topography in pen, pencil and with India ink, and colours."[46]

A key figure in modifying military science to include civil engineering after 1830 was Lt. Dennis Hart Mahan. Mahan had been a cadet at the academy from 1820 to 1824 and was one of the first to be subject to the new curriculum that taught a "complete system" of *la grande guerre*. After graduating, he spent four years in France, traveling and studying civil and military engineering at the esteemed School of Application for Artillery and Engineers at Metz. He then returned to live at West Point. In 1832, Mahan resigned his commission to assume the role of head of the Engineering Department. He was soon recognized for his expertise in civil and military engineering and for his dedication to the diffusion of military science, first to cadets and instructors at West Point, but then more broadly to civilians and militia officers through his writing. He became head of the Academic Board in 1838, a job he retained until his death in 1871. As head of the Engineering Department, Mahan was responsible for educating the First Class cadets in civil and military engineering, fortifications, and strategy (the science of war). In this capacity, he instructed all regular army officer graduates between 1832 and the Civil War, hence his influence on them is worth considering.

Very early in his teaching career Mahan became aware that the requirements for proper instruction of necessary subjects and the hours available during a four-year course of study were in fragile balance. A desire to add interesting or indeed important things to his civil engineering course would require reduction of time dedicated to other subjects. Reductions could easily lead to a diluting of subject material. Guided by the principle that fewer subjects taught with thoroughness were of greater benefit than more subjects dealt with lightly, Mahan was hesitant to expand the curriculum. In his mind, foremost in importance was the mastery of basic principles, through the in-depth study of particular elements. Only intense study could ensure that a subject was so thoroughly acquired that its application in any variety of circumstances was ensured. Mastery of theory gained by comprehensive learning of a specific subject would thereafter allow the graduate to draw on that knowledge in any future circumstance and adapt it to the situation, regardless of where these officers served.

One way that Mahan found to increase the scope of instruction without loss of depth was to reduce reliance on strictly European texts and ideas and to Americanize his subjects. He began to write notes that reduced the broad and complex European systems to just those things relevant to American defense and matters of internal improvements. He believed strongly that, unlike European armies, all regular officers of the U.S. Army required some civil engineering skill, even if not employed within the corps of engineers. He understood how easy it was to confuse cadets with complex civil engineering theory. Therefore, using his notes in lieu of European books, he taught the basics of this subject in a manner that an American officer could grasp. He also sought to furnish them with a text that they could use after leaving West Point. He published his lithographed course notes in 1837 in a book entitled *An Elementary Course in Civil Engineering*. This work became the standard text in many American colleges, went through three editions, and sold fifteen thousand copies in America, England, and India.[47]

Mahan's civil engineering class, conducted in the autumn term of the fourth year consisted of materials, masonry, foundations, walls and arches, carpentry (framing), road, bridge, canal, and railroad construction, waterway improvements, advanced theory in architecture, and, beginning in 1840, machines and industrial drawing.[48] His work was coordinated with the course in drawing, and he held high standards for the production of architectural plans. His instruction also included building military roads and architectural design for the West. In fact, graduates used his elementary course text (along with his 1836 work on field fortifications) in planning and constructing western military posts.[49]

Some historians have suggested that Mahan's civil engineering focus indoctrinated cadets in mechanistic thinking and that he forbade use of "inductive reasoning" in favor of "instantaneous, unquestioning obedience to a multitude of minutely detailed rules."[50] The broader inference here is that the emphasis on rigorous instruction of civil engineering not only took time away from the study of war but also inhibited cadet ability in military decision making because it enforced linear problem solving that restricted cadet creativity. These same historians have also castigated Mahan as being too theoretical in his instruction.

The authority on Mahan, historian Thomas Griess, disagreed. Griess's detailed study of the professor exposed how he and his colleagues remained acutely aware that a balance was required between theoretical instruction and practice.[51] Although Mahan had very little practical experience himself, he was Jeffersonian in spirit and believed in the practical application of all learned material. His course was not—as some historians maintain—"pure theory," but sought to teach how to apply theory in variances of circumstance. Mahan believed that flexible application could only come from a solid theoretical foundation, once stating that "the only really practical man is the one who is thoroughly grounded in his theory."[52] Unable to allow each cadet to construct actual things, Mahan made them instead draft mathematical models to demonstrate the application of theory. He also instituted class visits to internal improvement and fortification sites to allow cadets to see actual works.

Rather than encourage rote learning through memorization and recitation, Mahan desired that cadets apply basic principles to each problem. He called such application "common sense" and became renowned for its encouragement.[53] Mahan maintained a French philosophical point of view that it was possible to acquire perfect knowledge of a thing and that one could not acquire this through experience alone.[54] Perfect knowledge came when someone used a deep understanding of standard theory and guiding principles with common sense to determine "what is the proper thing to be done under given circumstances."[55] Mahan taught cadets how to think as much as what to think. His was a course in cognitive problem solving, which relied on use of creativity in applying principles. One historian has observed that this "common approach to technical problems" may have encouraged the emergence of a professional culture among all cadets, standards of problem solving being one of "the forces that bind an occupational community."[56]

Late twentieth-century autopsies on antebellum West Point often fail to reveal the broader social and military realities within which the academy functioned.[57] They portray the academy's narrow curriculum—containing fewer subjects taught in-depth—as a cult of engineering caught up in minutia for purposes of self-aggrandizement of the corps of engineers and unrelated to national defense or national need. Historians also criticize the practice of employing the most academically gifted officers in the corps of engineers, with one stating, "The wisdom of concentrating the available intellectual talent in branches of the army that would play minor roles in a war is questionable."[58]

In this criticism, there is a failure to understand the general-purpose role of the regular army in the antebellum period. Civil engineering constituted a significant portion of the academy's teaching because this skill was essential to an officer corps committed to developing the nation's Third System and in internal improvements infrastructure. The basis for such criticism dissolves when historians realize that the curriculum was attempting to produce some degree of expertise that graduates could use in a wide variety of improvement tasks, ensuring that officers of all arms could contribute to the strategic defense initiatives as well as help to ameliorate the primitive living conditions of the western frontier. More than half of all graduates in the 1820s and 1830s served at some time on topographical, engineer, or quartermaster duties, where they applied their West Point knowledge of civil engineering and surveying.[59] It is folly to think that these officers should have shed this work in favor of spending their time speculating about the need for larger standing armies or Prussian style general staff systems designed to plan grand campaigns of open offensive maneuver in Napoleonic fashion. The U.S. Army of this era served the federal government and not a supreme generalissimo, and their raison-d'être was federal policy, not military planning independent of political authority.

European peace allowed the United States to focus inward during the 1820s. The migrations of people westward and rising commerce increased public demand for improved transportation infrastructure and created competing demands for engineering and survey skill. The War Department met these demands by creating a bureau of topographical engineering and a board of internal improvements and by increasing the number of officers of all branches working for these organizations and for the quartermaster general. A great portion of the army was involved in work on Third System fortifications and internal improvements. War Department coordination of this work was made easier because many of the members of the Board of Fortifications served also on the Board of Internal Improvements. This created a synergy in strategic thought about integrated national defense, which at the same time supported broader social purposes and caused a shift in emphasis of West Point military science toward civil engineering and surveying.

Jefferson's vision of a national academy of science was realized at West Point by 1830. However, the more that the academy steered toward fulfilling the demand for scientific knowledge in civil engineering, the less time it could devote to military science involving fortifications, artillery applications, logistics, or the science of strategic movement. Reconciling the need to prepare officers in these subjects, while at the same time meeting the demands of a changing national focus, became a key challenge in the Jacksonian era.

5

Jacksonian Military Science

In the 1820s American military science expanded to include elements of civil engineering needed for internal improvements. During the 1830s and 1840s new demands led to further modifications. While the Third System continued to be the centerpiece of defense policy, the army became heavily burdened by Indian wars and removals in the course of western expansion. And while military and political leaders were still dedicated to preparing the nation for war against a European foe, some regular army officers questioned the continued validity of permanent fortifications and the Atlantic focus of defense policy. Advocates of steam technology wanted to modify the policy to include a role for railways and steam-powered floating batteries. Western oriented officers championed the establishment of cavalry. At the same time, a spirit of antiprofessionalism in certain quarters of Jacksonian America ensured that the regular army remained small and that West Point remained under constant pressure and scrutiny. Despite the demands and necessary changes to the academy curriculum, military science remained the core of instruction, vigorously defended by military leaders. During this period, Dennis Hart Mahan refined teaching of strategy and began publishing works to deliver military science to militia officers.

Antiprofessionalism and the Military Establishment

Antiprofessionalism and a desire to increase states' rights characterized the Jacksonian era (1829–60). State representatives to Congress frequently questioned the need for military appropriations for federally controlled military establishments. The West Point academy—the institution that most typified emerging professionalism—remained a target. However, Jacksonian Americans were not antimilitary. There was a general adoration of military pomp as an expression of state, sectional, and even national power, and people worshipped war heroes. Andrew Jackson himself personified a public sentiment of real military prowess in its amateur frontiersmen, not in its regular army. Jacksonians preferred romantic notions of natural martial genius and the innate fighting talent of Cincinnatus over any professional caste who hung their status on the peg of a hard-to-acquire military science.[1] Such sentiment led to occasional attacks on the military academy in Congress and in the press, the constitutionality of federal military establishments being in question. Occasionally, too, politicians or citizens accused the academy of elitism.[2] However, the attacks were relatively weak and in each case invoked a passionate statement of defense by one or more congressional representatives, by members of the executive branch, and by military and academy members themselves. On no occasion were annual appropriations for the running of the academy denied.[3]

Successful defense of the military establishment had much to do with the disposition of Andrew Jackson himself. While inclined to believe in a natural American fighting spirit, the U.S. Military Academy impressed Jackson, and he favored having a core of professional officers in the military establishment. He had seen firsthand in the War of 1812 and on operations in Florida the inadequacies of the militia.[4] He gave public praise to West Point in two annual addresses to Congress, ensured continued federal support of the academy, and remained an admirer of the works of Professor Mahan.[5] However, with regard to the regular army as a whole, Jackson remained a political man. In the finest Moderate Whig tradition, he protected the small standing army but refused to increase its overall strength.[6]

In equal measure, the successful defense of the military establishment can be attributed to the continued inherent logic of the Third System and its associated program of internal improvement. An alternative policy, requiring equal coherence, would have taken at least as much effort as that which had occupied Secretary of War Calhoun, Brig. Gen. Simon Bernard, and Colonel McCree between 1817 and 1822. Instability in the office of secretary of war during the Jacksonian era may well have contributed to the fact that no alternative policy emerged. There were nine secretaries between 1828 and 1848. During the same period, there were only two commanding generals of the army, Maj. Gen. Alexander Macomb (an engineer), 1828–41, and Winfield Scott, 1841–61. During the same period, there were only two chiefs of the corps of engineers, Col. Charles Gratiot, 1828–38, and Col. Joseph Totten, 1838–64.[7] All four of these officers were committed to the Third System, and the impact of their continuity cannot be underestimated.

The requirements to garrison Third System infrastructure substantiated the continuance of the small regular army, but not its growth. The army's authorized strength remained at 6,000 until Indian wars forced augmentations (to approximately 8,000) in the mid-1830s. The size of the army thereafter rose and fell depending on the threat. Real possibility of war with Great Britain in 1840 produced a surge in strength to 11,169. However, by 1845 the strength was down to 8,349. The Mexican War brought the army strength up to 21,686, but by 1850 it was down to 10,315.[8] Its small size necessitated constant augmentation by state and national volunteers, who were largely supplanting the militia.

The tasks allocated to the army increased steadily during this same period. Companies of soldiers were scattered throughout the country. They constructed and garrisoned coastal defensive works and forts along the northern and western frontiers, conducted internal improvements in every state, implemented the frustrating policy of Indian removals, and fought Indian wars.

The need for a larger military establishment to meet these tasks became obvious in 1836. The army then employed 650 officers and 7,310 soldiers divided into two geographic departments. The Eastern Department, under command of Winfield Scott, contained thirty-two posts garrisoned by 3,100 men. However, that year the War Department stripped Scott's command of 2,000 soldiers for service in Florida. As tensions with Great Britain rose steadily during the following year, the threat to the coasts posed by the Royal Navy exposed the extreme shortfall of soldiers in coastal garrisons.[9]

The Fortifications Board had based the Third System on a scenario of war against a European power, specifically against Great Britain. In this scenario, the principal threat was the Royal Navy's amphibious capability, with its 700 ships, 22,000 seamen, and 10,000 marines (backed by the 88,000-strong regular army).[10] The British maintained bases at Halifax, Bermuda, and the Bahamas from which to launch raiding forces (up to 20,000 strong) against America's eastern and gulf coasts to capture a critical city or harbor or to control openings to internal waterways. Memory of the events of 1813 and 1814 shaped this concern.[11] Third System planning dealt with all possible permutations of this scenario along the entire coastline. In the plans, the role of permanent fortifications, regular army garrisons, and local militia was clear. Well-sited defensive works manned by regular troops would deny capture of critical places, forcing the British to land in remote areas, from which they would have to advance by land on the critical points. Local militia forces would slow down British land movements in a manner similar to that employed by the Minutemen during the British retreat from Lexington and Concord, while other militias were mobilized and sent to the threatened region.[12]

As Third System planning and construction continued—incorporating new variations of this scenario—the issue of Indian threats on western or southern frontiers received little attention until mid-decade. In 1836 War Department concerns shifted from maritime fortifications to war with the Seminole Indians of Florida. The first foray into Seminole territory by American forces under Maj. Francis Langhorne Dade ended with the massacre of Dade's entire command, and seven years of war followed, costing the government $20 million. During that period more than 40,000 regular, militia, and volunteer soldiers served in Florida, where 1,500 died.

At this same time, the Western Department under Brig. Gen. Edmund Gaines also came under heavy strain. The department contained more than twenty posts garrisoned by 2,400 men. They faced an estimated 70,000 Indian warriors west of the Mississippi. A commentary from West Point, from one Habitator Montium, published in the *Army and Navy Chronicle* in October 1837, observed that on the western frontier the small number of soldiers scattered in small posts ensured that the Native saw no sign of the real might of the United States. The commentary suggested that the nation needed 15,000 to 20,000 troops to stabilize the frontier and, by "awe," prevent frequent Indian wars.[13]

Between the perceived threats posed by the Royal Navy and the western Indians and the real pressure of the Seminole War, a large portion of the army was kept involved in "a great experiment" to move Indians from the eastern states to designated territories west of the Mississippi. The federal government had formulated the removal policy in the 1820s in a series of land exchange treaties requiring westward movement.[14] Many tribes, such as the Sac and Fox, resisted. The State of Illinois requested War Department support for their removal, and Jackson authorized it by signing the Indians Removal Act in May 1830. With federal support, Governor John Reynolds mobilized state volunteers in 1831. The ensuing Black Hawk War proved devastating to the Sac and the Fox people. In 1832, because of corruption in civilian contracting of removals, the War Department assigned to the army both the task of organizing Indian movement and the settlement of cleared lands. In 1836 three Creek bands resisted but were suppressed and removed by

11,000 regulars and volunteers. The Cherokee also attempted resistance but were likewise forcibly removed. In 1836 alone, the army moved 400 Seminoles, 16,900 Creek, and over 1,000 Potawataniacs to the west side of Mississippi.[15] Of the estimated 130,000 Indians who may have lived east of the Mississippi in 1820, perhaps 30,000 remained in 1845. For the small regular army, the work of removals was emotionally hard, a far cry from *la grande guerre* that they thought to be their *raison d'être*.[16]

The tasks and operations of the army in Florida, across the Appalachians, and west of the Mississippi did not arrest efforts to complete Third System construction and to assist in internal improvements. Supporting efforts included army erection of a national network of lighthouses and federal infrastructure in the national capital.[17] All of this activity taxed the army's staff, requiring great numbers of officers of the line to perform staff functions, a proportion that grew as the numbers of posts occupied by the army increased. Table 4 is a snapshot of officer employment in 1835, but it is indicative for the entire period of 1830 through 1860.

After assuming the duties of the secretary of war in March 1837, Joel Poinsett moved to expand the military establishment to meet all of these new demands. He desired to reform the state militia system by conscripting 200,000 citizen soldiers. He felt that a regular army of 12,000 should be augmented by 12,000 militiamen drawn from this conscripted force to serve anywhere in the union for up to four years. The suggestion was reasonable given that at any time during the mid- and late 1830s, over 20,000 state volunteers and local militia served on active duty.[18] However, New England representative Daniel Webster and the Whigs eviscerated the proposal in Congress, denouncing its militarism as an infringement on state control of militia.[19] Poinsett's proposals set off tirades from various state militias, who also criticized the military academy at West Point. Such attacks only succeeded in bringing the secretary of war, the chief engineer, and the faculty at West Point closer together. Unable to secure a larger standing army or to reform the militia, Poinsett came to appreciate more quickly the uniform standard of professionalism of the academy graduates and the value of their military science, believing that the quality of individuals must make up for the lack in numbers.

The Shift from the Atlantic Focus

Besides insufficient numbers of soldiers, one other thing was becoming obvious under the various pressures of 1836–37. According to Poinsett's predecessor, the "general plan of national defence" suffered "too strong an impulse from the trans-Atlantic recollection," and Poinsett did not disagree. The Third System had never adequately addressed the northern and western frontiers, and western settlers were complaining about this imbalance. Only worn tracks between stockaded outposts demarked the boundaries between Indian and settler. When wars came to the frontier—as with the Black Hawk and Seminole Wars—the only way to deal with these was to strip the eastern seaboard garrisons. The army was clearly too small for all of the tasks, and there was a need to review plans, priorities, and requirements.[20] Poinsett set realignment in motion with the survey and construction work on the Great Military Road. He estimated requiring 5,000 to 7,000 soldiers to garrison these posts to placate public demand for greater security.[21]

In Brig. Gen. Edmund Gaines, the long-standing commander of the Western Department, the frontier settlers had an ally. Gaines had been in the army since 1799 and had spent a good part of his service in the West. A long rival of Winfield Scott, Gaines provided balance to Scott's Atlantic orientation, preference for professionals over state volunteers, and respect for West Point. Where Scott looked to Europe for military inspiration, Gaines sought native models to refine frontier warfare. Serving together in Florida in 1836, Gaines complained of Scott's pedantry and desire for regulation: "We have loaded ourselves with French and English Books, uselessly ponderous Company and regimental Books, and indulged in sedentary habits, until we find ourselves . . . barely qualified to cope with a savage foe."[22]

Gaines also publically rebuked West Point's system of inculcating "habits of thought," stating that this "quiet and almost exclusively sedentary mode of living" sacrificed "the vigor of constitution necessary to a real hard-duty soldier, to the attainment of that literature of science, with the social habits and enjoyments more befitting a country gentlemen of affluent fortune than a thoroughbred soldier."[23] Gaines held a particular dislike for West Point professors, admonishing those "who have never seen the flash of an Enemy's Cannon—who have acquired distinction only in the mazes of French Books, with only that imperfect knowledge of the French language which is better adapted to the Quackery of Charlatans than to the common-sense science of war."[24]

Gaines was right in several respects. For instance, West Point teaching did not cover Indian fighting. Gaines's plans to protect the frontier from "savages sufficiently numerous" called for the conversion of frontier military posts into military schools, designed to instruct western youths who were "unable to obtain admission into the Military Academy at West Point." These "cadets" would be associated with each frontier militia regiment and would receive a four-year course of military instruction, making them in "no wise inferior to the graduates of West Point."[25] Gaines was also adamant that the War Department limit the employment of infantry and artillery to coastal duties and replace garrisons in the West with cavalry units.[26]

To facilitate organization of the western frontier and to meet Gaines's demands, Poinsett won congressional support for a slight increase in the military establishment, but not for frontier schools. Congress increased the strength of artillery units and created a second dragoon regiment for western service. Poinsett then sought the assistance of the army's topographical engineers to conduct western surveys for internal improvements designed to create a stronger frontier. In 1838 he restricted employment of regulars in state-sponsored internal improvements. He then separated the topographical bureau from the corps of engineers and established an independent corps of topographical engineers.[27] Where the bureau had only ten engineers, the new corps had thirty-six officers (drawn mainly from re-mustered West Point graduates from all branches of the army).

The War Department first gave the Topogs responsibility for river and harbor improvements on the Mississippi, Ohio, Missouri, Arkansas, and Red Rivers. Throughout the 1840s, Topogs also conducted surveys on the gulf coast, the Great Lakes, western military roads and railways, the Texas Boundary Survey, and the Northeast Boundary Survey. The surveys were an extension of military reconnaissance work described in the texts of Lallemand and Gay de Vernon. Using the standard format for reconnaissance reporting, comprising survey maps and a memoir, the Topog officers used their West Point military science to compile tangible products for the War Department.

Col. Stephen Kearny's 1846 survey from Fort Leavenworth to Santa Fe to San Diego, a total of 2,200 miles, is a superb example of a military reconnaissance with auxiliary information.[28] During the expedition, 1st Lt. William H. Emory, class of 1831, wrote his *Notes of a Military Reconnaissance*. Acting as an advance reconnaissance party and report writer, Emory adequately mapped and analyzed the territory to determine the best locations for frontier forts and roads, aiding army war planning. His maps were the best of the period and laid down clearly the possible boundary lines with Mexico. However, Emory was also a scientific officer associated with astronomers, mathematicians, geologists, and the American Association for the Advancement of Science. His report had scientific appendixes containing information of interest to scientific, commercial, and political communities.[29]

While Topogs concentrated the lion's share of their work in the West, Poinsett was considering modifying the Third System by revising the defensive plans along the boundaries with the Canadas and New Brunswick. The United States had been in dispute with Great Britain over the boundary line in Maine for some time, and the increase of British soldiers in the northern colonies during the 1837–38 rebellions in Upper and Lower Canada brought forward the possibility of war over this disputed line. The secretary's "Report on Plans for Protection of Northern and Eastern Boundary," prepared in October 1837 and delivered in January 1838, criticized Congress's lack of energy to complete the original Third System works in these areas, and added that other works were now required to face the new British threats.[30] The Eastern Department commander, Gen. Winfield Scott, agreed and reallocated forces to the northern borders. The Board of Engineers then produced a report outlining various options for defensive and offensive campaigns on that frontier, presenting to the secretary of war the best defensive lines and bases of operation.[31] They also organized reconnaissances and surveys of these borders, commencing in the fall of 1840 when Lt. Robert Knowles, an artillery officer stationed at West Point, received a commission to traverse the British frontier between Niagara and Halifax to gather information about British defensive networks and the state of the military establishment in the colonies. Knowles transmitted the report, written very much as prescribed in West Point doctrine, to Secretary Poinsett. The Topogs then used it to prepare detailed surveys and defensive and offensive plans for war along the northern borders.[32]

In preparing for the possibility of war on all frontiers, Secretary Poinsett and several military commanders questioned the continued need for permanent fortifications. General Gaines was a critic of masonry fortifications and called for revision of the Third System because advances in steam power and other technology were outpacing improvement in fortifications design. He had thought thus since at least 1823, when he prepared a report on defenses of New York and New Orleans, and he reiterated this in a report to the secretary of war in December 1826 after an inspection of western military defenses.[33] Gaines observed that most major defense works remained incomplete and that "an epoch is at hand, in which the art of war . . . [as] regards the attack and defence of seaports, has undergone an unparalleled revolution." He was enthralled more by the promise than the reality of the technologies of the 1820s. He believed, for instance, that European steam ships could easily cross the Atlantic and surprise American cites and harbors and that advances in armor and naval gunnery had made permanent fortifications useless. In April 1836, in a report to Congress, Gaines suggested a reorientation of defense policy away from permanent fortifications toward smaller, less expensive forts supported by "steam frigates" and floating batteries. He wanted the secretary to consider steam power as an alternative to masonry fortifications.[34]

For Gaines, steam technology represented a "cultivation of military science." He urged the War Department to keep pace with improvements made in Europe so that a potential adversary would not have the advantage of "superior knowledge" and leave Americans to suffer the happenings of the "last war" when they "were comparatively ignorant of the state of military science."[35] Gaines also wished for a larger investment in railways, claiming that infantry and artillery reinforcements moving by road to threatened locations would take months to arrive. Railways, he believed, would allow the "fighting men of [the] central and western States the inestimable privilege of flying with unprecedented certainty, celerity, and comfort, to any of our vulnerable seaports."[36] Gaines's ideas regarding railways were indeed forward-looking and deserved consideration. However, his proposals were sometimes far-fetched given the actual state of technology at the time. That he had little knowledge of military science himself made him more susceptible to flights of fancy regarding the promise of technology. However, Gaines's criticism did provide other officers, junior in rank, the courage to question Third System thinking, and engineer officers between 1840 and 1860 periodically criticized the policy. Yet criticism of permanent fortifications in favor of earthen works, floating steam batteries, and railways did not negate the fundamental premises of the Third System. The principal threat to the United States was from amphibious invasion by a European power, and the United States required defensible frontiers that would impede invasion long enough for a volunteer army to muster and crush the invading force.[37]

Criticism and Defense of the Academy

The Third System policy survived criticisms in the 1830s and 1840s, a period during which the West Point academy was also on the defensive against Jacksonian antiprofessionalism and romanticism. Attacks came incrementally. The Annual Report of the Board of Visitors of 1830 couched an early critique of the narrowness of the West Point curriculum this way:

> The art of war is, and ought to be, the grand object of attention. It naturally divides itself into three branches—engineering, artillery, and tactics. The theory of each is explained on mathematical principles, whether for attack or defence, in works or in the field.... The board, however, ... would suggest whether the great and almost exclusive attention devoted to military engineering and the science of fortification does not retrench what is due to the "art militaire" in its most comprehensive sense; that is to grand tactics, and what may be termed the strategy of war.[38]

The board then proceeded to recommend a course on grand tactics with lectures and a textbook and descriptions of "remarkable" battles.

The professors at West Point, and Mahan specifically, agreed. Mahan preferred to use historical examples to teach his principles. However, time had always been the issue. There remained only eight months of instruction to teach three subjects in the fourth year—civil engineering, military engineering, and science of war (January and June being examination months).[39] The question then remained, which discipline should be cut in order to include a more deliberate study of history (or other liberal arts)?[40] External observers found an easy target in mathematics. The 1830 Board of Visitors had cited this:

> Mathematics are indisputably the basis of military science, and we would by no means disparage a branch of study of such pre-eminent importance. But mathematics are not the alpha and omega of the art of war.... We doubt whether it is essential to the for-

mation of an efficient officer that he be able to solve every question on the equation of osculatory circles. . . . Let us not destroy talent in its other beautiful and useful forms, by clipping it into triangles and parallelograms. . . . In the conduct of war genius of the highest order frequently exists without that peculiar bent which leads to excellence in mathematics.[41]

Statements regarding martial genius reflect shifting sentiment away from Jeffersonian faith in Enlightenment science toward emerging popularity of romanticism. Indeed, romanticism was spreading across American society, allowing poetry, literary novels, and fine arts to capture public appreciation. From the romantic perspective, virtue, especially martial virtue, was accessible only in the study of liberal arts. The 1830 Board of Visitors suggested that law, literature, languages, history, geography, rhetoric, English composition, and classical languages should replace mathematics. "Greek may be superfluous, but following the example of English military schools, Latin, if known, ought not to be forgotten." The visitors also proposed to fill the library with "cheap editions of English, classical, historical, and miscellaneous works," the underlying assumption being that military science and military virtue were separate, the former impeding proper learning of the latter.[42] The 1837 report from the Board of Visitors found cadets could not receive a complete course "unless the whole science of war, including the elements of martial and international law," was introduced.[43] Even a board under Winfield Scott in 1843 questioned the continued emphasis on mathematics. That board recommended excluding "higher branches of Mathematics and kindred studies" in order to "include more matter of practical utility." Their report cited two reasons for this: that in certain branches of the military service, higher mathematics was of little use and that a limited capacity in mathematics in some otherwise excellent cadets lowered their position on merit lists.

Historians have capitalized on periodic criticism of the curriculum to suggest that the American public believed the academy was failing to provide the military education required in Jacksonian America. Huntington used these instances to dismiss the academy's teachings as narrow "technicism." William Skelton, in an otherwise superb book, claimed that because cadets spent twice as much time studying math, science, and engineering as all other subjects combined, the academy failed to elevate cadet thinking to a point where they could appreciate strategy and the "philosophy of war."[44] Matthew Moten sees West Point military science as "Francophilia," stating that engineering squeezed out the possibility of studies in the humanities and that a "conceptual elevation of military engineering hampered a fuller understanding of warfare." Moten believes that liberal arts instruction might have allowed Americans to embrace contemporary Prussian practices and beliefs, which in the purest military terms—void of political realities—appear superior. In complete disregard for Third System or internal improvements, historians have concluded that a lack of education in the humanities led to the unfortunate circumstance that "when the profession needed men to concentrate on high-level problems of military policy and strategy, few were equal to the task."[45] What is meant by this is that America missed the opportunity to create a Prussian-style general staff, a larger standing army, elite military colleges, conscripted reserves, and elaborate war plans for *la grande guerre* that could re-create Cannae against any foe. The West Point academy is here judged against the "high-level problems of military policy and strategy" of Europe, not the United States.

Justification of the West Point curriculum today need go no further than to repeat the contemporary responses by Chief Engineer Totten. No other officer during the antebellum period was better able to explain publicly the coherence of West Point military science and its relationship to defense policy. In 1844, in response to a series of questions raised by the Senate Committee on Retrenchment on behalf of Generals Gaines and Scott, Totten wrote a forty-page letter to the committee head, the Honorable James A. Black. In it he addressed all the regular criticisms of the academy, especially criticism regarding mathematics and the all-arms curriculum. This letter satisfied the committee and effectively neutralized the attacks by generals.

Totten first appealed to egalitarianism. While acknowledging that infantry and cavalry officers might not need so much mathematics, he felt that all officers were entitled to the same education and that, except for the upper section of the First Class, cadets were taught no more than that "laid down for all the students of the most respectable colleges."[46] Furthermore, Totten believed that mathematics enabled military problem solving, and that its instruction served to "exercise and discipline the reasoning faculties, and to introduce a system and habit of thought which prove of the highest value in the pursuit of any profession." In addition, he knew that mathematics were fundamental to the acquisition of common skills. Basic surveying, reconnaissance work, road and fort building, bridging, and field fortifications were essential to all branches and required mathematical knowledge. Totten reminded everyone that if the War Department offered an alternative (one eliminating the need for mathematics, for instance), "the mass will crowd through that door, because, with the greater number of young men, when ease is put in opposition to toil, the benefits of the latter, of indeterminate value and wrapt up in a distant future, will not be considered."[47]

This type of response seemed to address contemporary criticisms adequately. It reinforced many reports supporting the West Point method. The 1837 Board of Visitors, for instance, felt that algebra, advanced geometry, trigonometry, and calculus were necessary "to discipline the mind of the pupil into habits of patient thought, and to prepare them for many of the important branches of science." The board felt that these sciences and mathematics constituted an "indispensable part of the education of the accomplished citizen soldier."[48]

To explain further his reasoning to a generation of politicians who had not been part of the post–War of 1812 reforms, Totten reiterated the rationale behind the West Point system. "Do not lose sight," he insisted, "of the fundamental principles that should always regulate our peace establishment, namely, the principle of a skeleton force, small, but capable of sudden enlargement, and perfect, in order that this enlargement may be brought into efficiency at the earliest day."[49] Understanding that military virtue, especially physical courage, was the touchstone of the state volunteer, he attacked antiprofessionalism by reminding political leaders that the regular army, trained in military science, was the essential core of an expansible military establishment. It was West Pointers that the War Department would rely on and promote to staff and command positions in a larger mobilized army:

> The staff of the regiments, and the greatly augmented staff of the army, would be supplied, for the greater part, from the same source . . . men perfect as examples of good soldiership, and capable of instructing in the best manner all around them. Discipline, order, health, efficiency, would spring at once from the mass which otherwise, with men, officers, and staff, all equally ignorant, must remain for months, if not years, without any other indispensable quality than courage, and this more often leading them into trouble, than to Victory.[50]

Totten cataloged the advantages that West Point offered in its insistence that all cadets learn the basics of infantry, cavalry, artillery, and engineering. He observed that no other nation educated officers for all branches of service: "The size of European Armies forbids such a procedure with them; our principle of a small skeleton force, however, not only admits, but requires it, if we desire to profit most by that skeleton principle."[51] Foreigners had commented on this "superiority" in the American method.

Of equal importance was the fact that infantry and cavalry officers seldom remained with their branch for an entire career. Totten explained that "these corps are to supply their full proportion of the high offices of the Army both in war and peace, offices that call for all the knowledge that man can acquire."[52] West Point prepared the foundation for service in both line and general staff functions, service required for career advancement. At the time of writing his response to Black, the army employed 223 out of 733 officers on general staff and engineering duties, 157 of these borrowed from line units.[53] Totten understood that the education at West Point was the greatest facilitator of this versatility of employment. In one succinct paragraph, Totten explained the scope of instruction at the academy and challenged the assertion that it was narrow and irrelevant:

> [The cadet] has been drilled in all practical military exercises; he has served four years in the ranks; for a year as a non-commissioned officer, and for an equal period as an officer in each capacity performing in the field all the appropriate duties, as Infantry, as Light Infantry, as artillery mounted and on foot, and as Cavalry. He can himself drill a battalion of Infantry, a battery of Artillery, or a Squadron of Cavalry, with all the promptitude and exactness of a veteran, on the black board he can explain the manoeuvers of the brigades, divisions, and corps des Armies, he has spent nearly a year in actual camp, under tents, performing all duties pertaining thereto. His mind invigorated by constant exercise and stored with a fund of professional elements: his habits molded by the observance of order and discipline to a prompt, efficient, and cheerful execution of duty, his moral character elevated by the practice of virtuous habits and the high toned sentiment and honorable emulation in which he has participated, thus stands the graduated cadet about to receive the appointment of 2nd Lieutenant.[54]

Replying to the suggestion that the academy reduce its instruction to tactical matters and contract scientific education to civilian colleges, Totten replied, "The education given at West Point is not scientific merely, not literary, nor military, nor theoretical, nor practical. It is all these combined. . . . No other institution has the same object . . . none of them having any such profession in view."[55] He acknowledged and encouraged the growth of military schools in the United States because the states needed large militia forces, but he added that no state could design and sustain an institute equal to the military academy at West Point.[56]

Change and the West Point Curriculum

During the Jacksonian era, Colonel Totten and several secretaries of war defended West Point in Washington, allowing the academy's Academic Board to concentrate their efforts on needed reform. Conscious of broad challenges posed by antiprofessional and romantic sentiment and the need to remain relevant in the eyes of the public, the board considered curriculum change to incorporate the exigencies of western frontier service, to update technological instruction, and to refine the teaching of military strategy, while at the same time conserving the core of science.

In attempting to balance curriculum reform and at the same time retain West Point's distinctive purpose, the long service of Academic Board members granted consistency if not obstinacy. Professor Mahan remained as head of the Academic Board from 1838 until his death in 1871. The other members, for the most part good friends of Mahan, were also long serving. William H. C. Bartlett, head of the department of natural and experimental philosophy, served from 1834 to 1871. Albert E. Church headed the math department from 1837 to 1878. Drawing instructor Robert Weir served from 1834 to 1876. Claudius Berard was head of the French department from 1815 to 1843, and Jacob W. Bailey, head of the department of chemistry, mineralogy, and geology, served from 1838 until 1857. Such lengths of service contributed to tremendous power of conservation, undoubtedly reinforcing a natural resistance to change and preventing radical shifts in the focus of teaching.

The senior military member of the board was the superintendent. This position changed hands six times between 1833 and 1860, from Sylvanus Thayer to René de Russy in 1833, to Richard Delafield in 1838, and to Henry Brewerton in 1845, to Robert E. Lee in 1852, to John Banard in 1855, and back to Delafield in 1856. Thayer advised each successor to be careful not to dilute the scientific element of the curriculum too much.[57] Occasionally a superintendent tried to limit direct contact between board members and the chief engineer and the secretary of war, particularly when issues caused dissent among board members. Joseph Totten, however, made superintendents understand that he and his political masters wished to know the minds of individual board members, whose "character and attainments" warranted agency.[58] The influence of the Academic Board was therefore not small.

The desire to conserve the "Thayer system" while addressing the need for change made curriculum review the hardest chore of the Academic Board. Regulations allowed only 10 to 11 hours per day for academic recitations and study. Military instruction and duties occupied another 3 to 4 hours, leaving time to eat (1.5 hours) and less than 2 hours of personal time (with 7 hours assigned for sleep). The curriculum reforms of 1819 had filled all of the available hours in a day. Each department protected every minute of allocated time with jealousy, making any change—even small ones—a hard exercise.

The most pressing problem was finding time in the curriculum for mechanics (particularly instruction in steam technologies), chemistry and mineralogy (of growing importance to Topog officers surveying the new territories), practical engineering (needed by officers charged with military road and bridge construction), and the study of grand tactics and strategy. The faculty had squeezed all but the last of these subjects into the curriculum by 1840. They added steam engine and communications technology to the instruction of mechanics and more advanced mineralogy lessons. These caused a reduction in French instruction. However, this had no adverse effects because by 1840 English books (a mix of translations of the French treatises and new works by American authors) had replaced most of the French texts. The tactics curriculum added cavalry and light horse artillery drill and riding lessons once horses were brought to West Point in 1838. Cadets learned to ride gun carriage horses in what the army called "flying artillery" batteries, a capability well suited for western service.

Infantry foot drill and basic cannon drill were reduced to allow this adjustment, but the number of hours dedicated to these basic drills was still sufficient to ensure competency in all arms by the time a cadet graduated.

Mahan held responsibility for revamping his First Class courses of civil and military engineering and fortifications to find more time to teach strategy. He understood that internal improvements and western service required a minimum level of proficiency in civil engineering, and he was satisfied with the standards of theoretical instruction that he was achieving. Mahan's course in military engineering consisted of subcourses in field fortifications, in military communications (bridging and roads), and in permanent fortifications, and he was happy with the coverage of these subjects, except that cadets lacked opportunity to practice.

Mahan and his colleagues knew that the four years at West Point were, for most officers, the singular time in their military career when they would receive formal instruction in theory, and they understood the importance of theoretical foundations of knowledge. However, the faculty also wanted a proper balance between theoretical instruction and practical applications, at the very least using mathematical exercises to demonstrate a familiarity with the theory of fortifications design, but employ more practical exercises for military bridging or other field engineering tasks.[59] The Board of Visitors in 1836 observed the practical bent of West Point instruction: "The course for instruction is very perfect. The cadets are taught the *rationale* of their studies, to think for themselves, and to apply their scientific attainments to actual practice."[60]

Yet Totten and Mahan both wanted more, and they sanctioned the purchase and use of fortifications and architectural models and the actual construction of field defenses during cadet training.[61] In 1842 Mahan, Delafield, and Totten agreed to establish a separate department of practical engineering under a serving lieutenant of engineers. This department took over teaching of "military communications," which included outdoor exercises for expedient bridging techniques, road improvement, and field entrenchments.[62] Mahan's texts were the basis of instruction.

Mahan knew the danger of perpetuating a public perception that West Point military science was an elitist field of study for the regular army officer alone, so he sought to make a portion of what he taught accessible to militia or state volunteer officers. If national defense ultimately rested on the foundation of a mobilized citizen army, then the War Department needed to diffuse military science to some degree throughout the militia and volunteer officer corps. Mahan felt compelled to force two aspects of this science on the militia: field fortifications and reconnaissance work. In a letter to the secretary of war in 1840, he opined that all officers—regular and otherwise—should demonstrate knowledge of "the Means of Military Reconnaissance and a knowledge of the simple elements of field fortifications."[63]

The importance of field fortifications reflected a long-standing conviction of the army. General Gaines had summed it up thus in 1827: "In a state of war it often becomes the duty of officers of every arm and every grade to plan and construct for the purpose of immediate protection against an enemy. . . . This duty necessarily devolves on the immediate commandant there present, whether of cavalry, artillery, infantry, or riflemen."[64]

In 1836, Mahan published a *Treatise on Field Fortifications*. He dedicated the text to "the officers of the U.S. Militia" and attempted to spread the essential elements of this military science to "those whose province it falls to instruct, discipline, and lead [the militia] into action."[65] Mahan desired that the militia officer could "take [it] with him into the camp" and therefore had it printed as a small, tightly bound booklet. In the introduction, he put his subject into strategic perspective, stating that field fortifications were a daily occurrence on a campaign and that the militia needed to use them as "conservative means" to strengthen their innate capability and mask their weakness. Regulars had the advantage of discipline. They were no more courageous than militia, but regular discipline and "habitual training" gave them "shoulder- to-shoulder courage," which was an advantage in open combat. The breastwork, on the other hand, gave the militia "equilibrium," as proven at Bunker Hill and at New Orleans. In order to preserve the lives of militia soldiers, who were too valuable to lose in tactics that squandered their blood, Mahan recommended that all officers master field fortifications.

Mahan's text was not a drill manual or an outpouring of technical data. It sought to give the reader a theoretical if not philosophical understanding of how field fortifications (and reconnaissance work) enabled tactical success. However, it was mathematical in its descriptions. Soldiers used field works to allow them to produce effective fire from behind cover.[66] To do this, they needed to know how to estimate ranges and how terrain influenced weapons fire. Target practice was necessary using a variety of terrain to see how the hardness of earth and undulations caused or prevented ricochet and the influence of palisades, abbatis, and other obstacles. Officers had to train to control fire and estimate ranges in order to determine how to concentrate fire and when to deliver the best fire effect.[67]

After citing the key principles of weapon's fire from defensive works, Mahan described the works themselves. Entrenchments had to provide shelter from enemy fire, be an obstacle to the enemy's advance, and grant defensive flanking fire. All field works conformed to Vaubanian profiles, with trench, dirt rampart, glacis if possible, and obstacles that would break the enemy formation in front of the entrenchments. The shape of the work—the *trace*—was a standard polygonal form with angular bastion projections used since the mid-sixteenth century. This trace kept all lines of defensive works to 160 yards or less (effective musket range) before angling away, allowing defenders to sweep the front of an adjoining part of the field work with flanking fire.[68]

In determining how to plan and construct field fortifications, officers had to know the penetration depths of various calibers of shot into earth and clay and wood and how to construct revetments and defensive obstacles.[69] All of this was set down for the militia officer to absorb, with the primary understanding that the epitome of the art of combat was in the physical manipulation of terrain to achieve the maximum effects of concentration of fire. In chapter 14, Mahan described how to attack permanent and expedient defensive works and how to conduct a siege in classic Vauban style using zigzags and parallels.[70]

It would be incorrect to state that Mahan's treatise was just a regurgitation of Vauban. He was actually describing the state of the art in European doctrine in the nineteenth century, and he incorporated the latest ideas from France that called for the entrenchment of an army within one night.[71] For officers destined for western service he included plans for the construction of wooden stockades, blockhouses, and splinter-proof shelters.

Mahan dedicated a portion of his work to other aspects of campaigning that he felt were essential for militia officers. Chapter 12 covered military communications, concentrating on expedient floating, ferry, and trestle military bridges. Chapter 13 was dedicated to military reconnaissance. Here Mahan repeated the theory of Gay de Vernon and Lallemand.[72] Militia officers should appreciate the requirement to produce the two parts of a military reconnaissance—a map and a memoir that described the resources of a region and its allowance for troop movements. For an officer to make a proper reconnaissance he needed knowledge of surveying and trigonometry and the science of military movement, including bases, lines, and angles of operations. This knowledge, when combined with observation on "the culture and produce of the country" including commercial and manufacturing resources, would assist an officer in making the proper reconnaissance reports. Americans should also seek information from "carriers, wood-cutters, hunters, trappers and Indians."[73] While he dedicated *Field Fortifications* to the militia officer, Mahan was essentially cataloging the same thing that he used to instruct cadets eligible to be part of, or seconded to, the newly formed corps of topographical engineers. Topog reconnaissance reports produced in the 1840s reflected Mahan's theory.[74]

Mahan's *Field Fortifications* was not a runaway best seller, but various publishers eventually printed it in no less than six editions, putting over 10,000 copies into circulation before and during the Civil War.[75] Through this and his other publications and his course notes, historians can discern Mahan's shift in emphasis. He had accepted that the American system of mobilization, dependent on volunteer forces, placed limitations on the instruction of military science. He therefore desired to take complex European theories, written for professional soldiers, and simplify them in order to give American amateurs something they could apply in North America. The content of his engineering courses provided basic instruction in a small number of field works and relatively simple theoretical instruction in permanent fortifications, bridging, and reconnaissance work. The publication of *Field Fortifications* and later on his *Permanent Fortifications* reveal his satisfaction with these subjects. But Academic Board notes reveal unease with the teaching of strategy.[76]

Having given over time in the fourth year to practical engineering (in the new department of military engineering), finding time for the proper instruction of strategy was even more problematic. Contemporary European theories were too complex to teach to cadets and militia officers, and they lacked applicability to the vastness of the expanding and sparsely populated United States. During the 1830s and 1840s, Mahan struggled with ways to render European theories accessible and give Americans an applicable idea of strategy. For this reason, he ceased using Gay de Vernon and Capt. John O'Connor's summary in the early 1830s and replaced these with his own notes, which contained derivations. Strategy was a part of his short subcourse on "the science of war," taught to the First Class cadets in their second term. This course covered the composition of armies, the principles of strategy, campaign marches, orders of battle, the choice and properties of military positions, military reconnaissance and encampments, and the general arrangement and organization of frontiers.[77] He has left his ideas to us in a series of lithographic notebooks that he used for instruction between 1835 and 1860.

Mahan's notes define strategy as "the Art of directing masses on decisive points" and the art that "regulates the movements beyond the range of the enemy's cannon." He admitted to an intimate connection between strategy and tactics, but suggested that they were distinguished from each other as "that which separates the science from the art." In this, Mahan perceived a science of military movements, governed by principles "founded on the importance attached to certain points or topographical lines."[78] He favored the theories of Bülow and Archduke Charles and certain elements of Jomini. The greatest difference between Mahan and Jomini, whose influence in the United States in the 1830s was still limited to Capt. John O'Connor's summary in Gay de Vernon's treatise, was Mahan's refusal to embrace Jomini's weighted emphasis in lines of operations above all other elements of campaign design.[79] For Mahan, the first and most important point in planning a campaign was identifying topographical points of strategic value (and in this he follows Archduke Charles more than Jomini). He defined a "strategic point" as a place that gave a major advantage to an army, determined by the topography of a theater of war. Three kinds of strategic points were important: those which united the base of operations (defined, in accordance with Bülow, as "the line from which operations start"), those which are at the end of an operational movement (called the "objective points"), and points which were intermediate between the base and the objective.[80]

To be truly strategic, a point had to lie on or command a strategic line. The principal strategic lines were the *base of operations* and the *line of operations*. Mahan clearly expressed the essential nature of the relationship between these two. While twentieth-century historians too often focused on the role of lines of operations in discussing Napoleonic or antebellum military theory, contemporary military thinkers like Mahan gave equal attention to the base and its relationship to the line of advance, giving all discussions from this period a geometrical character. It was irrelevant if these points and lines were framed within Bülow's triangle or Jomini's quadrilateral depiction of a theater of war.

Mahan's first type of strategic line, the base of operations, comprised the principal strategic points on a frontier (fortresses, depots, communications hubs, etc.) necessary for the provisioning of an army for a campaign. A proper base of operations facilitated transportation between the interior depots and the frontier, and allowed offensive movement from it into the enemy's territory. Such movement would be on the second type of strategic line, the line of operation that an army followed between the base and its objective point. Once an army commenced its movement upon a line of operation, the line behind the army (connecting it to its base) became its "line of communication." Mahan agreed with Bülow that a base of operations should be long and rest on several points from which offensive marches could occur and thereby enhance the security of the possible lines of operations. He stated that if all the "necessaries for an army [were] collected in one depot the operations would be in daily danger from any accident happening to the single route on which the supplies are transmitted." He suggested that there should be an easy communication between all these strategic points of supply along the base and that it would be the greatest advantage to have each one fortified. The extent of the base of operations should be in proportion to that of the line of operations, in order to keep the enemy from operating on the flanks of the latter and by that means threatening the safety of the base. If the base of operations was insufficient, a second should be established as the line of operation was prolonged. This second or "accidental" base should be defensible, and once established, "it should hold the same relations with respect to the new line of operations, as the primitive base did to the old. These reciprocal relations are very much the same as those of the parallels of a siege to the [zigzag] approaches on the place."[81]

The *sin qua non* of this science of movements was concentration of force on an objective point, preferably on the flank of an enemy. In his brief description of the various types of combinations possible to achieve this—using different types of lines of operation—Mahan cited Jomini's preference for interior lines, but kept his

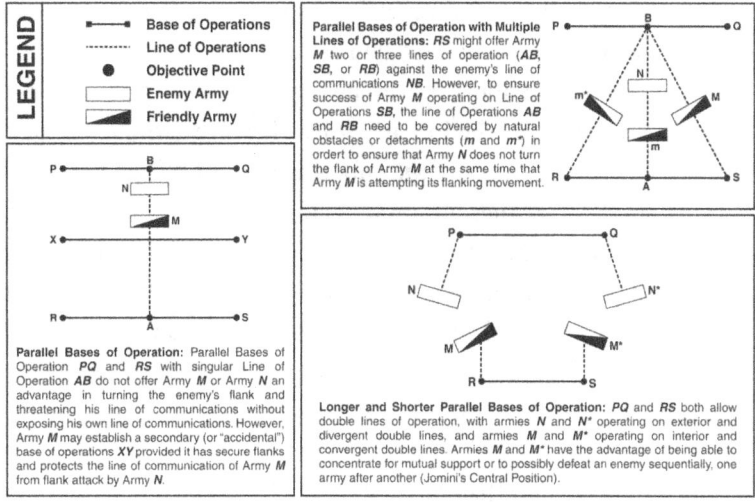

FIGURE 5. Mahan's Theory of Strategy, the Science of Movement, ca. 1840. There is no proof that Mahan used these specific diagrams to teach his cadets, as they do not appear in his lithographed notes. But Mahan's wording of "Composition of Armies" indicates an approximation of Dufour's work, with portions based on Archduke Charles and Jomini. The diagrams are the best contemporary representations of the simplified version of strategy that formed an alternative to the complex combinations of Jomini, Archduke Charles, and de Bulow. Based on "Principles of Strategy and Grand Tactics" by Dennis Hart Mahan in *Advanced-Guard, Out-Post, and Detachment Service of Troops: With the Essential Principles of Strategy and Grand Tactics for use of the Officers of the Militia and Volunteers* (New York: John Wiley, 1863), 174–175, plate 7, figs. 14–18. These were taken directly from the English edition of G. H. Dufour's *Strategy and Tactics* (New York: D. van Norstrand, 1864), which was first introduced in North America as *Tactique (Cours de)* (Paris, 1841).

balance by giving equal place to bases, lines, and points. He implied a need for flexibility of thought regarding the use of combinations to suit strategic circumstances. To Mahan, this was more about maintaining the initiative through celerity and concentration than about following a singular prescription. This distinguished him from O'Connor, who attempted simplicity by reducing Jomini to one preferred combination.

The object of all of this geometric reasoning was to use the "fundamental principle of war" to "gain an advantageous peace . . . by decisive strokes." In applying the principle, Mahan was adamant that an army "assume the initiative" by offensive actions, because to await attack was to play "the subordinate part." Using the element of surprise, armies attacked against weak points, especially those allowing attackers to gain the flank of an enemy or to fall on his lines of communications with his base of operations and thus threaten destruction.[82] In these prescriptions, Mahan emphasized activity and celerity and the need to concentrate and give battle and pursue a beaten army vigorously, seeking its complete defeat. In the application of the fundamental principle, the morale of the troops was no less important than "the physique," and officer courage was no less important than intellect. The "physique" was not simply a matter of massing forces but doing so on the best strategic points and lines to achieve superiority. Historians have unfortunately generalized such reasoning as "Jominian" and by associating it with Jomini himself castigated the whole system of thought.

A revision of history, beginning with J. D. Hittle's edition of Jomini's *Art of War* in 1947 and gaining traction under Russell Weigley's compelling pen, claims that Jomini cast a powerful spell over antebellum military thinking. This view is unsustainable. There is a lack of evidence suggesting that Jomini was widely read before the Civil War or that officers knew and debated his systems. Except to a few who could read his French texts, Jomini's pedantry with a multiplicity of lines of operations was unknown to the U.S. public during much of the Jacksonian era. While several translations of his 1804 *Treatise on Grand Operations* (which Capt. John O'Connor as used in his appendix to Gay de Vernon) were published in England after the Napoleonic Wars and may have been purchased by Americans, Jomini's *Summary on the Art of War* (1838) only became available in one translated limited edition in the United States 1854. The Civil War sparked wider interest and a new edition in 1863. The academy used his *Summary* for one experimental year (1859–60), but discontinued it. Beyond this, Jomini was not well known.

What Mahan taught, and therefore what cadets and graduates of the West Point academy knew, was a very simple and more balanced concept of campaign design that gave more emphasis to the relationship between the base and the line than between comparisons of interior versus exterior lines. This does not mean that Mahan can escape the label of being "Jominian." It does, however, call into question that label. Mahan and most other military writers of the period maintained geometrical explanations of campaign design simply because they embraced Enlightenment thinking and they lived in an era where topography defined the limitations of movement and concentration of armies. Topographical features were best portrayed on maps that defined the key points and lines (mountain passes, transport lines and hubs, water courses, and natural drifts) that would shape the advance or retreat of armies and define the number of combinations between armies. It was a simple step to transfer map points and lines into geometric figures. The inclination of officers toward geometric understanding of strategic combinations and battlefield evolutions reflected Enlightenment faith in mathematical explanation of the universe, not some theoretical hegemony by Antoine Jomini.

Mahan reduced more complex European descriptions and combinations of Bülow, Jomini, and Archduke Charles into short but precise lessons on the science of movements he called strategy and on the related subjects of marches, subsistence, castrametation, convoys, and detachments, all components of logistics. His instruction was traditional, emphasizing the importance of well-established depots for provisions in a base of operations, secondary magazines established on lines of operations, and provisional magazines "in the immediate neighbourhood of an army containing supplies for 8 or 10 days." Mahan believed in "the necessary subordination of the military operations to the measures of supplies." He acknowledged the need to sometimes rely on local requisitions, but felt that total reliance on foraging was suitable only as a temporary measure "in a thickly populated and well cultivated country," which did not describe most of the United States.[83]

Mahan's instruction of strategy within his course on the science of war was based on his thirty-two-page handwritten lithographed notebook and was delivered in six recitations, each three hours in length. With associated evening study periods of three hours and morning study periods of an hour each, the academy devoted 24 hours of study and 18 hours of recitation to this subject (see tables 1, 2, and 3). An additional four recitations (12 hours) and study periods (16 hours) were dedicated to a review of this material, bringing the total to 30 recitation and 40 study hours in this topic. This increased in the late 1850s. The clarity of Mahan's short work on strategy, the method of learning of the academy, and the fact that most of the principles incorporated within the course were the same as covered within instruction in field and permanent fortifications suggest that students mastered Mahan's simple theoretical prescriptions and retained more of this subject than previously admitted to by historians.[84]

Mahan and others had long struggled with ways to include historical examples to reinforce the principles taught in these recitations. His short course on the science of war was insufficient in and of itself to delve deeply into historical analysis that could render the general principles more accessible. Adding more to this course would mean displacing instruction in civil and military engineering. From the mid-1830s onward the idea of adding another year to the curriculum circulated, allowing thought for an expansion of topics. Mahan supported a recommendation to introduce some liberal arts, and in 1838 he had put forward a resolution to consider establishing ways to do this, including teaching history to the Third Class.[85] However, while agreeing in principle, he was unprepared to endorse more short ancillary courses of study that only served to promote superficial memorization of "a chronological table."[86] Mahan accepted memorization for recitation in subjects where material memorized was concrete and would lead to application, such as mathematics and physical science, but refused to condone memorization of material of no applicability other than pedantry.

Chief Engineer Totten, Superintendent Delafield, and Mahan all began to consider adding a fifth year of instruction at West Point to allow them to adequately address deficiencies in the curriculum: "The portion of the course which is now most defective is that relating to the Science of War, in which are comprehended Strategy, and its kindred subjects."[87] At the same time, they considered a program of advanced study for officers, either as students or for those returning to West Point as instructors.[88] This was logical. The military academy relied on officers borrowed from line units for one or more years to act as assistant instructors. Between 1802 and 1861, 305 graduates returned to teach at the academy. Capitalizing on this, an advanced seminar on the study of military campaigns was started for instructors in 1848 (discussed in chapter 6), and a graduate course for engineer officers began in 1844 (see chapter 7).

In 1843, General Winfield Scott chaired a special Board of Visitors of nine military officers who formally recommended a five-year program, including more "literary studies."[89] Totten agreed with Scott's findings and directed that the Academic Board study the proposal.[90] Mahan chaired the committee and responded with an endorsement.[91] Totten then directed Delafield to proceed.[92] Subsequent Academic Board minutes reveal that Mahan wanted more time for the science of war, including twenty-six recitations. He admitted to having cut some of the course during the 1830s in order to achieve a more thorough rendering of civil engineering and fortifications theory. A fifth year would allow time for an expanded course in strategy and the study of military history without jeopardy to other subjects. His 1845 committee studied Scott's proposal that cadets retain works on history in their rooms to study when time allowed during the five years. However, the committee felt that this would divert cadet attention away from the study of term courses.[93] The committee did support allowing time for study of English grammar and composition because these subjects had utility to the officer. It was not in favor of the study of literature or metaphysics, however, because they perceived the utility of these subjects in military affairs to be marginal.[94] The war with Mexico disrupted deliberations over the five-year program, and it took another nine years before board members agreed about the program's content.

On the verge of the war with Mexico, the War Department and army leadership were largely content with the program, content, and product of the military academy. Americans continued to question the department's Atlantic perspective as threats and requirements on inland frontiers emerged. Regular officers questioned the Third System permanent fortifications and pushed for new technologies. The academy introduced new weapons and tactics, particularly those involving the horse. Field engineering was evolving, too, and railway construction and steamboat usage predicated rethinking in army logistics planning. Jacksonian militia officers asked for and acquired treatises on tactics, and regulars became committed to experimentation in new methods for field artillery and cavalry. In all this spirit of advancement and change, however, the military science of West Point remained intact, changing only by efforts of the professors to Americanize European texts and teaching. The stalwart walls of the academy resisted the current of romanticism that swept around it, ignored hot winds calling the institution unconstitutional and elitist, and presented itself as the bulwark of professionalism even when many Jacksonians found the idea of professionalism abhorrent. Criticisms of West Point would only diminish sharply after the war with Mexico when the work of the regulars, and their military science, proved decisive.

6

Military Science during and after the Mexican War

Graduates of West Point applied all aspects of their military science in the war with Mexico, something unacknowledged in the historiography. The consensus sees the Mexican War as a training ground for future Civil War commanders, reinforcing the school of belief that experience is the only teacher in war and that experience mixed with natural talent, not study, makes the great commander. This reading backward of history ignores the application of doctrine and attributes none of the success in Mexico to military science. Political and military leaders acknowledged the contribution of West Point graduates at war's end, a point overshadowed by contemporary tensions over proper command and control of the military establishment. The Mexican War had revealed the power of the Washington general staff bureaus and the resentment this caused among commanding generals. The war also revealed the importance of higher military education and affected how West Point considered teaching strategy.

War Policy and Strategy

In 1844 James Polk ran for election on a platform of expansionism, and upon winning the presidency he moved to fulfill his promise of "re-occupation of Oregon and the re-annexation of Texas."[1] His strategy involved a series of military expeditions to seize northern Mexico and California and force the Mexicans to agree to new boundaries. He annexed Texas in 1845 and ordered Brig. Gen. Zachary Taylor to advance to the Rio Grande in March 1846 and secure the river as the new border with Mexico.[2] That June Polk directed Col. Stephen W. Kearny to depart Leavenworth, Kansas, march 850 miles to capture Santa Fe, and then to proceed to San Diego, annexing northern Mexico and southern California.[3] Fighting commenced on the Rio Grande in May, and in August Taylor marched into Mexico to seize Monterrey. Polk then ordered Brig. Gen. John Wool from San Antonio to reinforce Taylor. When it became clear that the Mexicans would not cede the disputed territories without war, Polk ordered Brig. Alexander Doniphan (following Kearney) to leave Santa Fe in December and join Taylor. Simultaneously Polk ordered Winfield Scott to capture Vera Cruz on the Gulf Coast and to advance on Mexico City to force its capitulation.[4]

None of these operations was a scenario envisioned in Third System defense policy. Yet all were conducted using prewar practices and West Point military science. From a strategic perspective, the war involved the establishment of military frontiers and bases of operation, reconnaissance and long marches over harsh unmapped terrain on various lines of operation, a conventional siege, and complex combinations in the face of a regular Mexican army superior in numbers.

To prosecute the war, the United States mobilized 104,000 soldiers: 60,000 state volunteers, 32,000 regulars, and 12,000 militia troops.[5] Polk appointed thirteen volunteer Democrat politicians as general officers for these forces, in what was an attempt to counterbalance the Whig leanings of Generals Taylor and Scott. But at no time did any of these appointed generals outrank or command the two regulars.[6] Secretary of War William Marcy and the regular army ran the war. Exercising a system of centralized control distinct from the 1812 experience, the office of the adjutant general (Brig. Gen. Roger Jones), acting directly under the secretary, corresponded with General Taylor and state governors to muster the appropriate forces. Quartermaster General Thomas Jesup coordinated the logistics requirements of first Taylor's then Scott's armies, using a decentralized contracting system to respond to each of the general's clothing and equipment needs. Regular army assistant quartermaster officers labored constantly to find transport, eventually managing the purchase of more than 11,000 horses, 22,000 mules, 16,000 oxen, and 6,800 wagons.[7] The chief of ordnance, Brig. Gen. George Bomford, supplied their arms from existing arsenals and from manufacturers.[8] The Springfield Armory produced 14,200 muskets, and the Harpers Ferry Armory produced 12,000 stands of arms between 1 July 1846 and 30 June 1847.

Almost all of the officers in the staff departments were graduates. However, West Pointers also served in line units. John Calhoun's plan for an expansible army was enacted, and "skeleton regiments"—which had officers but only half of the required privates—received the remaining soldiers. About three-quarters of line officers of the regular regiments were academy graduates. New regiments were added to the regular army's order of battle, each containing a core of graduate-trained officers. While most of these regiments were unaccustomed to large-scale maneuvers, "the theory of larger operations had been carefully studied by a majority of the officers." To assist in implementing this "theory of larger operations," each army had a complement of West Point–trained engineers and Topogs.[9]

The president and war secretary directed Taylor to make maximum use of his attached topographical engineers in the advance to the Rio Grande.[10] Captains George Meade and Thomas Cram and Lieutenants Jacob Blake and Thomas Wood made a reconnaissance of the two hundred mile route from Corpus Christi to Matamoros, laid out camps, and reported on distances, availability of water, wood, and grass, and road and route conditions. Taylor's chief of staff, Capt. William W. S. Bliss, then faced the challenge of how best to supply the advancing army.

Bliss was a man knowledgeable in military science. Having graduated from the academy in 1833 into the infantry, he returned to West Point as an assistant instructor of mathematics from 1834 to

1840. He then become chief of staff to the commanding general in Florida, 1840–41, assistant adjutant general to the First and Second Military Departments, and finally chief of staff for Taylor in 1845. Bliss had not been avoiding line duty during this period, but was employed in staff functions because of his "marked mathematical ability" and his strength in "arranging military movements, drawing up orders, and conducting voluminous correspondence."[11] He was tasked to move the army from Corpus Christi to the Rio Grande.

Working with unit quartermasters and the attached engineers, Bliss organized an "experimental column" to traverse the route and prove it capable of sustaining the thousands of soldiers, ordnance, wagons, and animals that would make up the rest of the army.[12] He also dispatched ships to carry Maj. John Munroe, engineer Captain Sanders, officers of the ordnance and pay departments, a field battery, and a siege train to occupy a base on the north end of Brazos Island, at Point Isabel, a strategic point for landing, for a supply depot, and for an encampment. Point Isabel subsequently served as a base of operations for land and naval forces at the eastern end of the new frontier on the Rio Grande and the northern end of a maritime frontier on the east coast of Mexico (see map 5).

Marching to Matamoros were the likes of 2nd Lt. Ulysses Grant of the Fourth Infantry, class of 1843. Grant declared later in his memoirs that he had not taken to studies at West Point, claiming instead that his time there was "devoted to novels."[13] As unlikely as this is, Grant did find mathematics "very easy," and he contemplated becoming a mathematics teacher.[14] Forced to the frontier as an infantry officer after graduating low in his class, Grant wrote to Professor Church in 1845 asking to be assigned as assistant professor, and at Fort Jefferson he continued his studies in mathematics and "valuable historical works" with "regularity, if not persistency."[15] His inclination toward math made him an easy choice for staff duties, and in the practice of the time, the army seconded him from company work to become assistant quartermaster of the Fourth Infantry.

On 25 March 1846 the advance party of Taylor's army reached

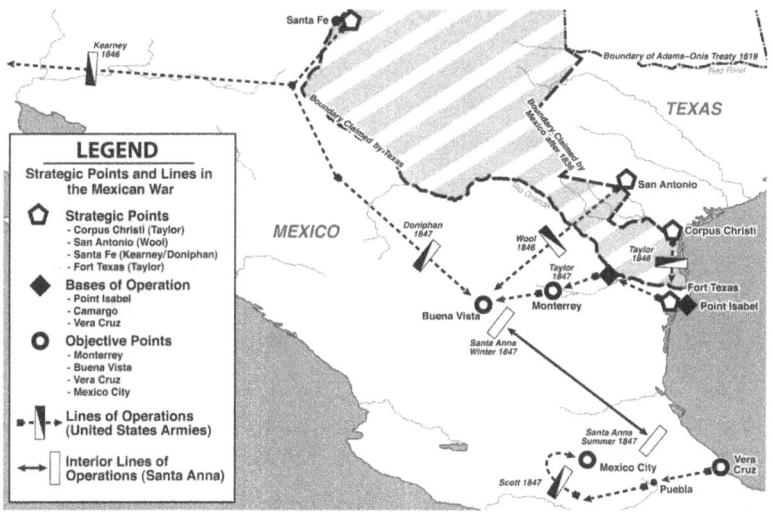

MAP 5. Strategic lines and points of the Mexican War. Based on a map from *Guns Along the Rio Grande, Palo Alto and Resaca de la Palma* by John S. Brown, at http://www.history.army.mil/ (accessed March 18, 2011; not included as part of copyright).

Point Isabel, where it established a fortified supply depot at the mouth of the Rio Grande. A small garrison was left at Point Isabel while the bulk of Taylor's troops continued to advance up the Rio Grande to a point on the river across from Matamoros. There an advance column under Maj. Jacob Brown was busy preparing field fortifications. Capt. Joseph Mansfield, class of 1822, fortified a camp for five hundred soldiers with the assistance of Capt. Edgar Hawkins of the Seventh Infantry (class of 1820) and Capt. Braxton Bragg of the light artillery (class of 1837).

Major Brown built Fort Texas (later called Fort Brown) to Mahan's specifications. It dominated the Rio Grande with three of six bastions commanding the town of Matamoros opposite. Its perimeter was eight hundred yards and contained a rampart nine feet high, fifteen feet thick, with a ditch eight feet deep.[16] The remainder of Taylor's force encamped behind expedient defensive works to the north of this new strategic point. Happy that Fort Texas would fix the Mexicans at Matamoros, Taylor returned to Point Isabel. The Mexicans promptly attacked the fort between 3 and 9 May 1846. Whilst outgunned and outnumbered, the fort's walls and design easily withstood the assaults. Taylor moved quickly to relieve Fort Texas, engaging the enemy at Palo Alta and Resaca de la Palma en route on 8 and 9 May 1846. He fought these actions in the open without field fortifications but not without military science. The superior training and handling of the "flying artillery" by West Point officers was decisive, as it was to be in many subsequent engagements.[17]

Taylor then seized the initiative by invading the Mexican province of Nuevo Leon and marching on the objective point of Monterrey. He tasked Williams and Meade, commanding a mixed force of pioneers, infantry, and cavalry to conduct reconnaissance and repair the road ahead of his column. This force quickly invested the town. Taylor sought the advice of his regular infantry, artillery, and engineer officers to formulate his plans and attacked various forts around Monterrey on 21 September.[18] The town capitulated three days later, and with its surrender Taylor secured an eight-week armistice, during which he ordered Meade to complete detailed reconnaissance reports for the new military frontier.[19]

While Taylor was investing Monterrey, Brig. Gen. John Wool advanced a second army from San Antonio to the Rio Grande. Just as Taylor had done, Wool employed his attached Topogs to perform a military reconnaissance to Presidio and Parras in Mexico. The language of their report, using the terminology of Mahan's course on the science of war, described the best base and line of operation for Wool's advance.[20] Wool began his march in September 1846. He employed his chief engineer, Bvt. Lt. Col. Robert E. Lee, to repair roads along the chosen route. Upon reaching the Rio Grande, on 12 October Wool's army crossed on a "flying bridge," a prefabricated pontoon design that had been moved by the engineers from San Antonio.[21] Wool then moved into Mexico to link up with General Taylor.

In the meantime, President Polk, furious at Taylor for his unauthorized armistice, feared a loss of momentum and ordered volunteer reinforcements by sea to Point Isabel and Fort Brown and to a new base of operations at Camargo, where army quartermasters and commissariat officers were busy stockpiling supplies for advances further into Mexico. Volunteer Brig. Gen. Robert Patterson then advanced with another column from Fort Brown to join Taylor in January 1847. The newly created regular army Company of Sappers, Miners, and Pontoniers led his advance, under Capt. Alexander J. Swift, protégé of Mahan, graduate of the French engineer school at Metz, and instructor of practical engineering at the academy. His fellow West Point instructors Gustavus Smith and George McClellan were his second and third officers.[22] Both of these junior officers left accounts of the subsequent campaigns, revealing their use of military science.[23]

The first task of the new engineer company was to repair roads from Matamoros to Victoria and Buena Vista to allow more rapid movement of forces on this new frontier.[24] McClellan was happy for this work as it could lead to combat and, with combat, an opportunity for glory. He exposed his farsighted ambition and his contempt for the prospects of building Third System fortifications when he stated: "I am (I hope and believe) pretty well cured of castle building. I came down here with high hopes, with pleasing anticipations of distinction, of being in hard fought battles and acquiring a name and reputation as a stepping stone to a still greater eminence in some future and greater war."[25]

Generals Patterson, Wool, and Taylor concentrated their forces at an objective point at Buena Vista in early 1847. Taylor had hoped for permission to conduct an advance on Tampico or even Mexico City with the combined forces. But the overland line of operations from a single point base at Camargo would be tenuous. The Mexicans could easily outflank it. President Polk instead ordered an amphibious landing at Vera Cruz under General Scott to open a second front. All regular infantry and engineers departed Buena Vista in January to join Scott's army, staging out of the Point Isabel base. Taylor kept his topographical engineers, some artillery and key staff, and his volunteers. The Mexican general, Santa Anna, took advantage of Taylor's weakened forces and attacked Buena Vista in February.

During the ensuing battle, Taylor's Topogs distinguished themselves in command of line units vindicating West Point's all-arms focus, and West Point graduate Braxton Bragg acquitted himself notably in command of his artillery. Once Santa Anna retreated, Taylor consolidated his forces on the frontier, with the Topogs writing reports on the military requirements for Texas and New Mexico.[26] The position at Buena Vista became an alternative base of operations to the one Scott was establishing at Vera Cruz—something that Santa Anna could not ignore, forcing him to consider threats from the north as well as the east. This use of alternative bases of operations in a single theater of war containing one objective point (Mexico City) might have defied Jomini's ideas of the central position and preference for interior lines of operations, but this was no matter for these American commanders, whose ideas of strategy were more flexible than Jomini's.

Vera Cruz to Mexico City

General Scott's army landed at Vera Cruz on 9 March 1847, besieging the city and forcing its capitulation on 28 March. While Scott deserved much credit for the success of this short siege, he was not without support. According to one young officer, "The capture of Vera Cruz was an affair, in the main, of the staff and artillery."[27] This staff included engineers, who according to T. Harry Williams "constituted a kind of unofficial General Staff."[28] Immediately after landing, eleven engineer officers under Chief Engineer Joseph Totten and Robert E. Lee began nine days of constant reconnaissance work to determine "points of attack" on Vera Cruz. Lieutenants Beauregard, McClellan, Smith, and Foster sited artillery positions and worked with artillery officers to construct gun platforms and ammunition magazines. Second Lt. Cadmus Wilcox, class of 1846, was proud of how "the engineers located and constructed the batteries with such good judgment and care that there were few casualties." Meanwhile, Scott's ordnance officers disembarked the siege train, and his quartermasters offloaded and stockpiled supplies. Engineers and artillerymen built a road between the landing beaches and the proposed siege lines and organized the movement of heavy siege cannon and mortars and ammunition to the battery sites. Wilcox claimed "the fixed ammunition used by the artillery was prepared under the direction of ordnance officers with a skill insured by their education and their experiments and labors of laboratory."[29] Siege trenches were opened on 18 March.[30]

On the 19th, Scott's army started digging Vaubanian parallels, as taught at the academy. The artillery positions were set in *grande place des armes* and the guns commenced firing on 22 March. So well sited were some batteries that the Mexicans did not know where they had been constructed until cannons opened fire.[31] The Mexicans then fired hundreds of rounds at these batteries and trenches in what became an artillery duel. In this the Americans employed ten mortars, three 68 pounders, three 32 pounders, four 24 pounders, and two 8-inch howitzers. The Mexicans understood enough about the science of siege craft to know that the fall of Vera Cruz was inevitable under the fire of such guns and the advance of zigzags from the parallels. They surrendered on 29 March. Military science had spared the need for a costly infantry assault on the Mexican fortifications. That day Scott wrote that victory was delivered by his "indefatigable Engineers."[32]

Fear of the arrival of the yellow fever season compelled Scott to move inland in April 1847, in a drive to capture Mexico City using the single mountainous road leading west as his line, with Vera Cruz as a base of operations. He set out with his 10,700 soldiers and found that Santa Anna's forces had blocked his way at the first defendable hill, Cerro Gordo. Scott immediately sent out engineer reconnaissance parties to define the terrain and the enemy's positions, the first of many such sorties during the upcoming campaign. In his postwar report, the chief topographical engineer described the value of engineers in the reconnaissance role:

Accurate geographical and topographical knowledge of a country are particularly essential to military operations. They are the eyes of the commanding general. With these he can see the country, and can know how to direct and combine all his movements or marches, whether offensive or defensive, and without them he is literally groping in the dark, incapable of devising plans for his own operations, or of anticipating those of the enemy. With this knowledge, war becomes a science, in which intellect will ever predominate over numbers; without it war becomes a mere exhibition of physical force: slow, expensive, and often disastrous, as numbers and courage can only be relied upon.[33]

Robert E. Lee and Lieutenants Beauregard and Tower gained distinction at Cerro Gordo for discovering and reporting a route around the north flank of Santa Anna's forces. During the ensuing battle on 18 April 1847, Lee and Tower and Ordnance Lieutenants Hagner and Laidley demonstrated the value of their all-arms training by commanding artillery from heights overlooking an exposed flank of the Mexican army.[34]

Scott repeated this successful outflanking maneuver numerous times during the march on Mexico City. After seizing Puebla in May, he rested his soldiers and built up his supplies. The last portion of this march was conducted without attachment by supply convoy to his base at Vera Cruz, a bold decision taken to allow his soldiers to move with celerity and to maximize the number of soldiers in his attack columns (not having to detach soldiers to guard his line of communications).

Scott's engineers led the advance to Mexico City, finding ways to circumvent Santa Anna's defenses (containing 30,000 soldiers) east of the city. In Scott's words, "the boldest reconnaissance of the war" occurred during the night of 17–18 August when Lee and Beauregard wove their way through the Pedregal lava beds to find a path to the San Angel road and outflank Santa Anna to the south.[35] Lee's engineers marked the route, then improved the trail and guided the army around this flank, forcing Santa Anna to retreat into the city.

The engineers repeated this tactic at Chapultepec, with a reconnaissance party leading the army south and west of the city, which was then captured by an audacious infantry attack supported by field artillery. In mid-September Santa Anna conceded defeat, and Scott's army entered Mexico City. In his official report to the secretary of war, Scott praised dozens of officers, all regulars and West Point graduates, for their performance during the campaign. His commendations were not just for engineers but included assistant adjutants general, assistant quartermasters, chiefs of staff, ordnance officers, and aides-de-camp. Scott concluded that "the capital, however, was not taken by any one or two gallant corps, but by the talent, the science, the gallantry, the prowess of this entire army."[36]

This army recognized the role of the engineers. It was known to many that Scott never made a decision about an attack without conferring with them about "the decisive point." They provided him with *coup d'œil*.[37] Grant felt that "both the strategy and tactics displayed by General Scott in these various engagements . . . were faultless." But he also praised specifically the engineer officers, "who made the reconnaissances and led the different commands to their destinations, . . . so perfect that the chief was able to give his orders to his various subordinates with all the precision he could use on an ordinary march."[38] The only maneuver and attack in the entire war that was unassisted by preliminary engineer reconnaissance was General Worth's "impetuous" assault at Churubusco, which proved to be the costliest assault of the war.[39]

But the success of the West Point graduates was tempered by their losses. In all, 523 graduates served in Mexico; 73 dragoons, 153 artillery, 189 infantry, and 108 as key staff (18 engineers, 22 Topogs, 16 ordnance, 11 assistant adjutant generals, 29 quartermasters, 8 paymasters, 4 subsistence officers). In addition 36 retired graduates led volunteer units. Of the 523 regular army graduates, 141 (27%) became casualties, of which 49 died. In December 1848

Secretary Marcy granted to the U.S. Military Academy the "appropriate depository" of the trophies of the war in honor of the graduates whose fortifications, artillery, ordnance, and logistic knowledge were so evident and whose engineering and reconnaissance skills provided the necessary information for planning and decision making. The Mexican War pretty much ended talk of abolishing the academy, probably because "there was no failure in the undertaking of any military operation or expedition during the war resulting from a want of education in the graduate."[40]

Back at West Point Professor Mahan followed the war as best he could in national newspapers. A biographical note of Mahan, written by one of his friends, later claimed the following:

> When his teachings bore fruit upon the fields of Mexico, enthusiasm broke through his habitual reserve; the success of his pupils to him was the vindication of their alma mater against the clamors which had filled the air, and, as one of the officers expressed to me: "He welcomed me home with moist eyes and a silent grasp of the hand more eloquent than words."[41]

The Aftermath

Beyond praise of West Point, one immediate result of the Mexican War was a tension between the commanding generals of the main armies, Winfield Scott and Zachary Taylor, and the staff bureaus in Washington.[42] Often late in submitting requisitions, Taylor nevertheless complained about lack of supplies, as did Scott. The quality of supplies was also a source of complaint, with generals implying that the quartermaster officers chose the cheapest contractors to fill orders because of more concern for fiscal efficiency than military effectiveness. These were old complaints commonly made by commanders in the field against staffs who lived comfortably away from the fighting. The reality of the Mexican War was that things ran rather smoothly given the problems of geographic distances between the source of supplies and the army, the harsh weather and terrain, and the vast size of the operational theater. The singular biggest failing was in the poor health conditions and the want of hospitals. As with all wars before World War II, the majority of American deaths in Mexico were from disease.

While complimentary of West Point graduates acting in the capacity of staff with the field army, Scott believed that the permanent staff in Washington wielded an uncomfortable amount of bureaucratic power. The permanent staff bureaus were not new, having been created during and after the War of 1812. But throughout the intervening decades their powers grew, much to the frustration of commanders such as Generals Gaines, Macomb, and Scott. Some historians maintain that this reflected a clear and dysfunctional division in the antebellum officer corps between staff and line, an argument that exposes divergent thinking about what constitutes a proper military organization.[43] Historians have found it easy to castigate the separate staff structure under political control in favor of one under a single commanding general (following the Prussian model). There are few histories of the Washington general staff bureaus and therefore few champions of their work. The preponderance of histories of this period focus on general officers or the exploits of the army in the West. This imbalance in the historiography makes it more difficult to appreciate the work of the general staff and the corresponding importance of a doctrine based on military science used by the staff.

The main source of evidence for staff and line tension lies in the records of complaints of the generals in chief of the army and several geographic department commanders. Gaines, Jackson, and Scott had all criticized the independence of the bureaus under the secretary of war. They had little authority in these complaints. Legislation in 1818 had clearly placed the bureaus under a political master and granted no legislation to recognize powers of a commanding general. Major General Macomb had court-martialed Adjutant General Roger Jones in 1830 for insubordination, hoping that the War Department would recognize his position as general in chief as a necessary link in the chain of command between Jones and the secretary of war. While he succeeded in this court-martial, he failed to stop the practice of the staff bureaus reporting directly to the secretary.[44] After he returned from Mexico as general in chief in 1848, Scott started a long and very public campaign against the War Department over command and control of the army. In disgust at not being able to control all military decision making, he eventually moved his offices from Washington to New York in 1853.

Scott's opinions were not totally ignored. Postwar fears that some of the permanent staffs in Washington were too deeply entrenched in the political bureaucracy led President Polk to complain of their independence. Secretary of War Jefferson Davies echoed this in 1854, suggesting the elimination of all permanent staff appointments and subsequent designation of line officers (infantry, artillery, and cavalry) to temporary service in the staff posts.[45] Mahan himself described the "Washington Bureaus" as "those miserable nests of petty intrigues."[46] Capt. Henry Halleck (a man thoroughly appreciative of the need for good staff) offered this sentiment:

> The subordinate officers of the staff of an army, in time of war, are charged with important and responsible duties connected with the execution of the orders of their respective chiefs. But in times of peace, they are too apt to degenerate into fourth-rate clerks of the Adjutant-general's department, and mere military dandies, employing their time in discussing the most unimportant and really contemptible points of military etiquette, or criticizing the letters and dispatches of superior officers, to see whether the wording of the report or the folding of the letter exactly corresponds to a particular regulation applicable to the case.[47]

Historians have made much of these tensions as it helps to champion the stereotypical line officer—a professional suffering low pay, lack of prospect for promotion, and isolated postings on the dangerous frontier—against the image of a fat, comfortable staff officer in Washington.[48] Besides, the general staff of the U.S. Army put greater emphasis on logistics than on the more historically appealing problem of war planning (the focus of the European general staffs). In such criticism there is too little consideration given to the role of a centralized staff in the steady implementation of defense policy, strategy, and doctrine or in the common application of regulations. Nor do historians accord sufficient consideration to the demand by civilian masters for accountability. The U.S. Army staff organization, with its inherent compartmentalization of power between staff bureaus that reported to the secretary of war and to Congress, as well as the avoidance of centralizing power under a commanding general, reflected the continued suspicion of military establishment by politicians and the people. It was illustrative of the general American preference for a distribution of power, accountability, and fear of the potential dictatorial power that might evolve under one supreme general. Complain as some generals did, this was the reality of the antebellum peacetime military establishment. After all, this was the United States, not Prussia.

The staff system remained largely unchanged during the antebellum period because the staff bureaus did not fail in their responsibilities to Congress for the proper planning of expenditures, disbursing of funds, and accountability for purchases—functions completely independent of the general in chief.[49] It is hard to isolate when, if ever, this system failed the U.S. Army in any true sense, beyond being the target of the normal petty rivalries between competing officers. The fact remains that throughout the antebellum era, the permanent staff bureaus executed their staff functions without precipitating any military disasters and to the satisfaction of Congress, despite frequent changes in administrations. Their work brought consistency and regulation to an army so widely dispersed that without them numerous small military provinces would have emerged, each assuming the characteristics of the local commander.[50] Without the central staff bureaus, no strategic planning or preparation for war would have been possible, and the army's conduct during the Mexican War would have been diminished.

Of greater impact on the army than rivalries between staff and line officers was the continued practice of drawing officers from the line to fulfill general staff functions for one to two years at a time. The Washington staff bureaus themselves remained small, with only a handful of officers and secretaries. Most of the staff officers worked within the small garrisons of the army. Each department commander had officers at headquarters performing assistant adjutant general, quartermaster, and paymaster duties, overseeing the administrative paperwork, supply, and pay of the department. These officers communicated with line officers fulfilling assistant adjutant general and quartermaster functions at each military post. Secondment of line officers left many regular companies with only one or two officers present to oversee discipline and curb drinking and desertion. Table 5 shows the numbers and percentages of officers employed in staff functions and the number of military posts occupied throughout the antebellum period.

Departmental quartermasters followed the direction of the Washington bureau and passed to them the requests of the departments and subordinate commands. Line officers, seconded to the quartermaster general bureau as district, regimental, or post assistant quartermasters, contracted civilian companies and directed the distribution and disbursement of funds and material for the building and repair of roads, bridges, barracks, and garrison buildings and the movement of soldiers and supplies. Paymasters circulated through posts to verify accounts and oversee payments. Engineers, both military and topographical, worked alongside line commanders but remained directly responsible to the chief of engineers or chief of the topographical engineers in Washington. However, during all years of the antebellum period, line officers were also seconded to these corps to conduct engineer tasks.

The manufacture and supply of military ordnance was also independent of line commanders. This function rested squarely in the hands of ordnance staff officers located at the nation's two weapons manufacturing arsenals and various storage depots scattered throughout the country, ready to apportion arms to departments as directed by the chief of ordnance. While by no means thoroughly adequate to the requirements of the United States, the system was effective and was studied by the British as a model for the development of their own national ordnance system.[51]

Thus West Point graduates working as staff officers made this staff system function, despite tensions in Washington. While some generals complained of a division between line and staff, the reality in the military posts throughout the United States was that line and staff functions were performed by officers well known to each other from the same units, the majority being West Point graduates.[52] Their common education and cohesion made the system work. They shared a paradigm of military preparedness for the war. The evidence suggests that throughout this period administration, supply, armaments, munitions, and fiscal accountability remained satisfactory to the outposts, the War Department, and Congress alike.[53]

Officers employed in staff functions required staff skill and knowledge. Those working in staff bureaus required some skill in mathematics, while those in the ordnance bureau required knowledge of physical, chemical, and experimental sciences. The topographical bureau required officers skilled in astronomy, surveying, mineralogy, botany, and even ethnographic knowledge. Because the topographical bureau was also the repository for war plans, officers working there required knowledge of the science of strategic movements. Those working in the engineer bureau needed some knowledge of all of the above as well as skill in engineering science. Before, during, and after the Mexican War, it was West Point graduates who brought adequate knowledge to these staff positions. This might have put pressure on the academy to emphasize the importance of staff duties more rigorously in the curriculum, had not the euphoria that followed the victory in Mexico elevated notions of military glory instead.

The Mexican War had little impact on the curriculum at West Point, save to affirm its content. Returning officers did not demand any changes in West Point teaching as a result of their war experience. The war did, however, elevate ideas of military grandeur. While notions of personal glory never had a place in the academy's curriculum (there being no study of novels or hagiographic historical texts), West Point after the war resonated with a sense of promise to young men desirous of seeking personal acclaim in the pursuit of arms. This sentiment reinforced the prevalent spirit of romanticism that had been sweeping throughout the United States for over a decade.[54]

The distaste for novels at West Point that reflected the Enlightenment worldview had started to wane in the 1830s. The regulation forbidding the keeping of literature other than military or scientific texts was dropped in 1836, and the academy library began to carry and loan all of the popular literature of the day. Library circulation records for cadets and faculty show the popularity of Walter Scott, Tennyson, Coleridge, and Carlyle as well as American favorites James Fenimore Cooper and Nathaniel Hawthorne. Literature, particularly that which portrayed glory in military exploits, appealed to the young men who marched the West Point plain in fancy parade uniforms and who woke, walked, and went to bed to the sound of military bands.

Society's love of the novel, romantic poetry, and painting was creating an appreciation for art as a means to explain and enrich daily life and serve a function of moral instruction and inspiration.[55] So while the West Point Academy stood solid in its refusal to teach liberal arts, outside of the recitation rooms the plain of West Point was awash with sentimentality regarding the military courage, honor, and glory displayed in Mexico and the positive American mission perceived as manifest destiny.[56] The books and magazines, paintings, and music that were becoming an integral part of the cadet's life outside of the classroom reinforced these sentiments.

Even Professor Mahan, a man solidly of the Enlightenment tradition, enjoyed the escape of the novel, while his son, Alfred Thayer Mahan, had a passion for novels and hagiographical histories about the glory of Lord Nelson's navy. The elder Mahan understood all young men's reverence for role models of military courage and genius, but he abhorred the recurring idea among youth that genius was natural: "They magnify their idols, they love to regard them as something divine; they forget that human perfection is ever the result of long and severe labor."[57] Mahan was determined to show that real martial genius was also indebted to military science. For this reason, soon after the close of the war with Mexico, he agreed to chair a new society at West Point, the Napoleon Club.

In this postwar period, officers became interested in the Napoleonic Wars and the study of history. In France this stemmed from nostalgia. In the United States this interest thrived in a mix of romantic nostalgia for the American Revolution and a fascination regarding Napoleon's genius. A group of officers at West Point asked Mahan in 1848 to preside over an extracurricular association dedicated to the study of Napoleonic campaigns. Mahan was flattered by their invitation, stating that it was a rare thing that a schoolmaster should receive such an honor from former pupils.

The society met on Wednesday evenings in the academy building to hear papers presented by its members. Mahan asked that all who attended be "guarded and considerate" because the meetings would "doubtless originate discussions; interchanges of thoughts and opinions upon objects of interest to the Association; these will often prove earnest in expression." He therefore advised that "whatever may be said or done within this room . . . let it rest here. . . . Let no wrath nor bitterness go out from here with us."[58]

The Napoleon Club included voluntary members of the staff and faculty of the academy and was dedicated to "mutual improvement in the science of war."[59] Mahan expected members in turn to prepare papers regarding a particular campaign (usually Napoleonic or one of Frederick the Great's) and to present the papers before the association, invoking discussion and analysis. Papers were to be between thirty and sixty pages in length (but often much longer) and researched from a dozen or more sources available in the academy library.⁶⁰ To aid the club, Mahan had an 18' × 11' map painted on the wall of the meeting room depicting Europe as it was during Napoleon's reign as emperor. In 1862 President Lincoln was so impressed by this map that he asked for and received a copy.⁶¹

One of the club members recalled this as a highlight of his four years as an instructor:

> We had several clubs. . . . Best of all was the Napoleon Club. Professor Mahan was president of this, and gave out the Napoleon campaigns to be discussed by each member. Six weeks' time was allowed to prepare the paper. We had ample authorities, both French and English, at our disposal in the library, and worked diligently on our papers. The campaign of Waterloo, by Lieutenant B. S. Alexander, was considered one of the best discussions ever made of that notable defeat of Bonaparte. The campaign of Russia, by G. W. Smith, and of Wagram, by McClellan, showed marked ability. I believe something of this sort has been introduced into the course of study for the cadets.⁶²

The Napoleon Club would today constitute a graduate level seminar course. Mahan ensured that it was an academic pursuit aimed at professional enhancement, all for the broader purpose of preparing the officer corps of the U.S. Army for war. Of its known participants no less than sixteen became prominent commanders during the Civil War. While existing records do not definitively prove so, many of the 130 instructors at West Point between 1848 and 1861 may well have participated. Mess record and instructor lists of the military academy, combined with library circulation records of these officers, show a marked interest in military affairs and history by these academy instructors. Without a doubt, their formation at the academy, particularly after having had experience in Mexico, had nothing but a positive effect on their professional competence. As many as twenty-three of these officers would hold senior command appointments during the Civil War, and at least twenty held senior staff appointments (see table 6).⁶³

In his letter accepting the presidency of the club, Mahan explained its real purpose. He stated that the association would "ennoble the profession of its members"[64] and that their endeavors would show their intellect and character and prove to any European observer that "talent, in no line, is handed down from father to son." Praising their Mexican War experiences, Mahan warned the members to keep working hard in order to avoid the fate of the Prussian army in 1806 at Jena, whose officers had allowed themselves to lapse "in[to] ignorance and self presumption from past successes." He assured everyone that he was personally committed to keeping professional learning alive "through study of the great campaigns of great captains," because only by study can talent be made ready for war.

Graduates demonstrated all aspects of their military science during the war with Mexico. Military engineering, particularly with regard to field fortifications and siege trenches, military communications (expedient bridges and road building), topographical and military reconnaissance, surveying and map making, field and siege artillery applications, advanced logistics, and strategy—the science of movements—were the precise disciplines that differentiated the U.S. Army from its Mexican foe. There was no shortage of tactical competence or courage in the Mexican infantry and cavalry units. But they lacked professional artillery, engineer, and general staff officers of quality. In terms of military science, the Mexican army was found wanting when faced with the American armies employed in accordance with West Point teaching. The legacy of the war was a renewed confidence among the officers regarding the professionalism of the soldier of the regular army and an appreciation for antebellum focus on *la grande guerre.*

Postwar tensions did arise between line and staff officers; however, these were limited to the bickering between Washington bureau chiefs and commanding generals. Cadets at West Point felt the legacy of the war more outside of the classroom than within. Depictions of martial glory inspired them, reinforcing the romanticism of this era. The growing interest in Napoleon Bonaparte and his genius for war reinforced this sense of romanticism. Professor Mahan capitalized on this interest in the academy's new Napoleon Club, in which he sought to reveal the mystery of Napoleon's genius by unveiling that his system of war was wonderfully full of learned calculation. West Point's curriculum of military science survived the Mexican War but faced further challenges in the significant changes of the final decade of the antebellum era.

7

Antebellum Military Science

The decade before the Civil War was one of relative security for the West Point academy. Its purpose was confirmed in Mexico, and respect for its graduates steadily increased. Outside of the academy, tensions existed between the army and the navy concerning the relative value of ships over coastal fortifications. Army engineers argued about whether Vaubanian fortifications could withstand bombardment from improved ordnance.[1] The remainder of the army was absorbed with Indian wars and Indian policy and with policing a vastly expanded territory. Military science, however, still dominated professional military thinking, albeit with modifications. A new five-year program of study at the academy allowed professors to expand the scientific curriculum and deliver advanced courses on strategy and history. Comparison of the curriculum to those of European institutions found it to be largely satisfactory. Mahan and graduates of West Point also published works for the use of the volunteer militia. On the eve of the Civil War, West Point graduates made up 75 percent of the officer corps, and 35 percent were applying their knowledge of engineering and fortifications, artillery and ordnance, and logistics. Their work, combined with the dissemination of numerous publications written for military and public consumption, ensured that military science was a well-established paradigm by 1860.

Criticism of the Third System

After the war with Mexico, the War Department remained firmly committed to the Third System. Brigadier General Gaines had proposed alternatives before the Mexican War, and Maj. W. H. Chase, an engineer of the class of 1815, carried on his efforts. Chase was never convinced of the need for Vauban style fortifications and lobbied to have his own views heard in Congress. His example encouraged a small number of others to challenge the stranglehold that Third System thinking had over defense policy.[2] In 1845 Capt. John Sanders openly questioned its eastern focus and urged that steamboats and floating batteries replace defenses on the Gulf of Mexico.[3] Another engineer officer, James St. Clair Norton, publicly challenged the Third System throughout the 1850s.[4]

Chief Engineer Totten rebutted the challenges himself and encouraged other officers to publicly defend the Third System. Capt. Henry Halleck, class of 1839, became one such champion. Totten used Halleck's considerable talents as a lecturer and writer to provide speeches and reports on the advantages of permanent fortifications. As part of this effort, Halleck published the first truly American treatise on strategy and policy. Drawing heavily on what Mahan had taught him and on European sources, Halleck in 1846 wrote the highly instructive *Elements of Military Art and Science*. This became one of the most important books of its day, making Halleck's reputation as a military intellect. The work was the clearest statement of America's military requirements yet written. However, because no publisher has produced an edition since the Civil War, twentieth-century historians have largely ignored it. When referred to, they mostly castigate it, simply because Jomini's ideas dominate several chapters, overshadowing the remainder of a book dedicated to a logical and reasonable explanation of American military needs. Its content was strategic and did not touch upon the mechanics of battlefield evolutions. The book is therefore important because it demonstrates the way that influential senior officers in the military establishment viewed war in the last decade of the antebellum period, and it exposes the depth and quality of military thinking in America.

In his *Elements*, Halleck identified five important branches to military art and science: strategy, fortifications or engineering, logistics, military polity (policy), and tactics. Strategy was the "art of directing masses on decisive points, or the hostile movements of armies beyond the range of cannon."[5] Engineering provided both knowledge of the use of fortifications to withstand attack and how to overcome obstacles in offensive operations. Logistics embraced the practical details of moving and supplying armies. Tactics was the art of bringing troops into action within range of cannon.

For Halleck, strategy regarded the theater of war rather than battle. It involved the selection of important points in the theater, their fortification, the lines of communications between these points, and the plan for operations along these lines. In a sentiment that clearly meant the empowerment of the general staff, Halleck stated that in campaign planning the chief engineer, and not the commander, decided on the plan for obstacles and obstacle crossing and that the quartermaster general decided how to organize support and movement of armies upon chosen lines.[6]

Halleck spent a chapter on bases and lines of operation and their characteristics and types (parallel, oblique, perpendicular, concentric, etc.), giving definitions and general descriptions using historical examples. Here he reiterated the campaign designs of Jomini, Bülow, and Archduke Charles, clearly enunciating his preference for Jomini and particularly Jomini's emphasis of interior lines.[7] For this singular portion, historians have labeled him Jominian, and indeed, he was with regard to campaign theory. However, so were most Europeans at this time. Jomini's influence reached its zenith in Europe in the 1840s and 1850s, when interest in the Napoleonic Wars held sway over military studies. Jomini was still relatively unknown in the United States. Two of the most prominent European theorists, Marshal Marmont of France and Englishman Patrick Macdougall, shamelessly repeated Jomini's interior lines argument in their own texts.⁸ Therefore, Halleck's regard for strategic lines of operation was, if anything, state of the art. Topographical factors still governed land warfare in 1850. The *sin qua non* of campaign planning was a polygonal rendering of the topography of a theater of war, wherein one used convergent lines following roads, waterways, and natural lines of drift to concentrate forces at a strategic point. Only two European theorists, French officer Auguste Frederic Lendy and Swiss Chief of the General Staff Guillaume-Henri Dufour, differed from the standard reiteration of Jomini by refusing to place more emphasis on interior lines than on bases of operation.⁹

To ensure that his descriptions were not totally abstract, Halleck in chapter 8 applied his theory of campaign design to a scenario of war on the northern frontier. Here he suggested that the best line of operation against the British North American colonies was through Lake Champlain and the Richelieu River to Montreal.¹⁰ This would have satisfied Napoleon and Jomini, it being the shortest line to a point that would sever British east-west lines of communications and one that would put the offensive force in a central and interior position to the enemy.

Halleck then outlined the strategical importance of fortifications. He advised a balanced appreciation for both open warfare and the use of frontier fortifications, and he demonstrated his preference for the systems of Vauban.¹¹

Halleck's next strategic topic was logistics: incorporating the fitting out of troops for campaigns, regulating marches and convoys, transportation, magazines, hospitals, munitions, supplies, camps, and cantonments. He proposed maxims applicable to all campaigns, the first being the establishment of frequent supply magazines along the lines of operation, levying requisitions or foraging only when necessary. Magazines gave an army "velocity and impetuosity."¹² The second set of maxims outlined planning data for offensive campaigns. Here Halleck described Napoleon's system, with soldiers carrying ten days' rations followed by a corps baggage train of 480 subsistence wagons carrying another ten days' rations (for 40,000 men). Depots and supply magazines were also established on the line of operations so that an army was never more than 10 days' march from a base of supplies.[13]

Next, in what was probably his most important contribution, Halleck examined the complex issue of American military policy in a chapter entitled "Military Polity: The Means of National Defence." Strategic in scope, it began by discussing the key considerations: economic, social, and demographic. Here he considered the mobilization capacity of America's potential rivals—Britain, Russia, Austria, Prussia, and France—emphasizing that Europeans maintained 1/100th of their entire populations in arms during peacetime. Halleck acknowledged that the United States could not accept such a large standing army (with 25 million people, an army of 250,000 would be the European norm), but there could be no doubt about the need of a peacetime establishment sufficient for the defense of a large nation. He was convinced that a well-prepared army in 1812 would have captured Canada and that the Black Hawk War would have been preventable if only two regiments had been available at St. Louis in 1832.[14]

For Halleck, the peacetime establishment was essential because the United States could not rely entirely on militias. He repeated George Washington's criticisms that "short enlistments, and a mistaken dependence upon our militia, have been the origin of all our misfortunes, and the great accumulation of our debt." He used Washington's challenge of militia efficacy: "The militia come in, you cannot tell how; go, you cannot tell when; and act, you cannot tell where; consume your provisions, exhaust your stores, and leave you at last, at a critical moment."[15] The performance of the militia in the War of 1812 had been no better, and Halleck believed that the only guarantee of proper militia performance was in fortifications: "The experience of all other nations, as well as our own, has abundantly shown that a newly-raised force cannot cope, *in the open field*, with one subordinate and disciplined. Here *science* must determine the contest. . . . But when placed behind a breastwork, they even overrate their security. They can then coolly look upon the approaching columns, and . . . exert all their skill in the use of their weapons."[16]

Halleck's intent was not to convince the reader of the advantages of a large standing army but only to emphasize how America should mitigate against the inherent weaknesses of the militia system. He emphasized the uniqueness of each nation's defense requirements. England needed an army of 150,000 and a 700-ship navy. France required 350,000 soldiers and a navy of 350 vessels. The United States, with a long coastline, a long northern land frontier, and a vast western frontier, could expect only so much from its 7,590 soldiers and its 77 ships. Therefore, the answer to the nation's defense needs was exactly as foreseen in the 1820 concept of the Third System—an integrated network of permanent seacoast defenses and a plan for mobilization. Permanent fortifications were passive in nature, yet once garrisoned they could contribute to active operations. Once they were constructed, their maintenance required little expenditure. In peace, no citizens needed to man them. They were also of little threat to the democracy: "Of themselves they can never exert an influence corrupting to public morals, or dangerous to public liberty; but as the means of preserving peace, and as obstacles to an invader, their influence and power are immense."[17]

Halleck then gave a synopsis of the effectiveness and status of work on the Third System sites. Here he had political intent. He rejected what "some military writers" were calling "a new era in military science" with its assertion that improvements in naval technology had outpaced land defenses and that "floating forces" could be substitutes for permanent forts.[18] While not naming them, he was referring to Brigadier Gaines and Major Chase. He used historical examples to support his claim that permanent fortifications were superior to defensive naval forces, and he finalized his argument by using tables of expenditures to demonstrate the considerable expense of relying solely on naval power, compared with the lesser expense of relying upon fortifications.[19]

The remainder of Halleck's treatise covered the optimum organization and size of the peacetime establishment, descriptions of branch functions, and the role of commanders and staffs. On this last subject, Halleck displayed his modernity in emphasizing the need for special education for staff and command. He believed that the standard of education of engineer and artillery officers naturally inclined them to higher performance in staff functions.

Halleck felt that establishing the correct ratio of staff to line was also essential, and he compared current American practice with that of the contemporary French army in Algeria, which was also attempting to secure a vast frontier. The French in Algeria in 1844 maintained 5 higher staff officers, 112 administrative staff personnel, 61 artillerymen, and 48 engineers per 1,000 soldiers (with the remaining 774 divided between the infantry and cavalry). In comparison, the U.S. Army in 1846 maintained 2 staff officers, 20 administrative staff personnel, 310 artillery (an indication of the army's role in garrisoning fortifications), and 5 engineers per 1,000 (the remainder consisting of 513 infantry and 150 cavalry). Halleck used this comparison to show that an army serving in austere conditions, such as those found on the western frontier, needed a higher ratio of staff officers.[20] He failed to present the fact that a considerable minority of the infantry and cavalry officers within that American 1,000 were habitually performing staff duties.

Halleck highlighted the importance of staff, and particularly a chief of staff, when commanding leadership was weak. Quoting Jomini extensively, he stated that "a good staff has the advantage of being more durable than the genius of any single man; it not only remedies many evils, but it may safely be affirmed that it constitutes for the army the best of all safeguards."[21]

Halleck advised self-education to potential commanders. He divided general officers into three categories: *theorists*, who by study and reflection knew rules and maxims, *martinets*, who knew only the parade evolutions, and *practical men*, who were guided only by their experience. He opined that the uneducated general—the *mar-*

tinet or the practical man—would fall back on mechanical drills and be paralyzed by a new event, that the best generals needed at least an element of the theoretical, and that self-respect should bring a general to study so as not to "find himself constrained to *guess*, and not knowing how to be right by *system*."[22] Reading was important, and Halleck listed books that he considered essential to the basic formation of a theoretical understanding of war.[23]

In all of this, Halleck reinforced the existing strategic plan for the nation's defense and at the same time substantiated the continued need for professional military education at West Point. He also reiterated the gravity of the threat posed by European maritime forces to the United States. The threat was reiterated again by Brigadier General Totten in 1851, in the report to Congress entitled "On the Subject of National Defences."[24]

Totten opened this report by demonstrating the consistency of strategic thinking in Washington, referring to reports dating all the way back to the War of 1812.[25] He reasserted that the real threat to the nation came not from Indian uprisings upon the western frontier but from Europe. He enumerated four categories of threat: attacks on commerce and navigation at sea; raids on one or more points of the coast; attacks on principal commercial cities from the sea; and attacks on smaller towns of the coast. Specifically he reiterated the scenario of a British squadron of twenty to thirty sail-of-the-line and "war steamers," harbored at Halifax or Bermuda, able to land 20,000 soldiers on American soil, a force larger than could be matched by the U.S. Navy and one that would be a constant and present threat from Penobscot to Point Isabel. Totten maintained that coastal fortifications were essential to counter such a threat. Forts prevented seizure or bombardment of cities or ships in harbor or navy yards. They prevented passage of a river or channel leading to interior waterways. They forced an enemy to disembark in remote areas where any movement inland would be resisted step after step by patriotic American militiamen.[26]

In an obvious rebuttal to specific challenges, Totten stated that no

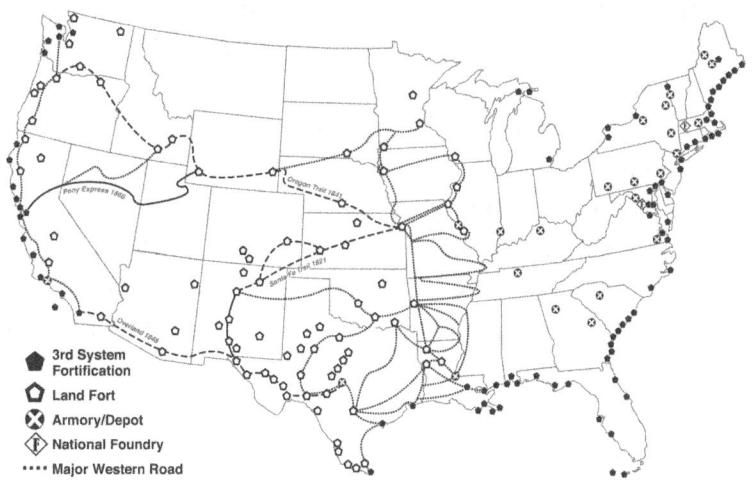

MAP 6. Third System sites and military posts, ca. 1855. Based on information from the Adjutant General's Office, Army Register (Washington: C. Alexander, 1850); and Joseph Totten, "Report on National Defense to the Secretary of War," quoted from House Document no. 206, May 10, 1840, 26th Congress, 1st Session; and Totten, "On the Subject of National Defences," 3–4 (Army Registers for 1844, 1848, 1850, 1853, and 1855). See also: [first name], Utley, [add title], [add pub info], 52–58; Ron Field, *Forts of the American Frontier 1820–91: Central and Northern Plains* (Oxford: Osprey, 2005), 5; Adjutant General's Office, Army Register for 1850 (Washington: C. Alexander, 1850).

system of railroads could replace Third System fortifications. Railways themselves could never prevent the capture of the vital waterways of Narragansett Road, Delaware breakwater harbor, Hampton Roads, Cumberland Sound, Key West, or the Tortugas. If the British seized any of these points, they would control maritime commerce of that entire region.[27] Beyond these points, no system of railway could cover all possible landing sites along the lengthy coastlines of the United States. Permanent fortifications therefore still provided the best defense of strategic points, with railways playing a subsidiary role of facilitating movement of mobilized militia to reinforce these points.

Gaines and Chase had already said much about using steam vessels and floating batteries in lieu of coastal fortifications. However, foreshadowing the battle between the ironclads *Merrimac* and the *Monitor* eleven years later, Totten explained that steam-on-steam vessel battles would inevitably create an equal fight, losing the advantage of well-sited artillery behind solid protective walls.[28] Totten's purpose here was to state unequivocally that national defense was still reliant on forts containing heavy artillery and that it was not feasible to rely on ships alone. Even England understood this, spending $1,300,694 between 1847 and 1850 to improve fortress defenses along her relatively short coastlines.[29] Economic statistics were always at Totten's fingertips. He cited the example of the fire of New York City in 1835 that had destroyed an estimated $17 million of property. Totten wondered how much greater would be the cost of allowing enemy vessels the ability to shell a city in a time of war.[30]

One may see Totten's resistance to replacing the Third System as self-serving for his engineer department. As head of the department and of the Fortifications and Internal Improvements Boards, he certainly maintained considerable monetary power and therefore purpose and esteem. But there were few if any coherent and realistic alternatives to this system, apart from increasing the size of the regular army, giving the War Department control of railways, or instituting a federalized active militia army in times of peace. None of these options was palatable to Congress during the antebellum period.

Although intransigent about strategy, Totten was not resistant to technological change. He was committed to improving the technology of coastal defenses, and he focused on improving casemate construction and designing embrasure configurations that allowed guns to swivel 60 degrees (as opposed to the European standard of 40 degrees).[31] He also sponsored advanced experiments in material resistance, in order to determine optimum composition of construction material, which led to the first use of armor plates on gun embrasures.

Totten's report outlined the status of the 186 coastal works that then constituted the Third System (see map 6). However, it also gave a summary of all the main strategic threats facing the United States. It was marked by its consistency of thinking and its reliance on military science. As the decade progressed, the use of the railway as a means of transport became more and more important, and sentiment grew for linking railways into the Third System. Since 1837 the federal government had loaned officers to railroad companies, and by the mid-1850s there was considerable knowledge resident in the army about rail use for moving supplies. Secretary of War Jefferson Davis in 1855 pushed to establish more railway lines, including one that would link the east and west coasts and expand Third System thinking to a truly continental level.[32]

Army Work, the West, and Military Doctrine

While military doctrine and policy after the Mexican War reinforced Third System principles and remained European in its inclination, the army focused on the territorial expansion in the west and implementation of Indian policy. These were complex undertakings. The demographic movement of settlers westward—accelerated by a Mormon mass movement into Utah and the California Gold Rush—outpaced the capacity of Washington to organize an effective Indian policy, leaving confrontation between settler and Native Indian inevitable. The army remained caught in the middle. It had increased in strength with the Mexican War to an authorized establishment of 12,927. But because of recruiting and desertion problems, the actual parade strength in 1850 was less than 10,350.[33]

Since 1838, Col. John Abert's topographical engineers had led western exploration. Between then and 1863, seventy-two officers served as permanent Topogs (all but eight were West Point graduates), publishing seventeen volumes of official reports on explorations, reading twenty-three papers to learned societies, and writing thirteen volumes of the *Pacific Railroad Reports*.[34] As officers of one of the army's scientific branches, many were connected with such scientific circles as the National Academy of Science and the Smithsonian, and their publications added greatly to general knowledge of six academic disciplines regarding the West: geography, geology, botany, zoology, ethnology, and archaeology. The Topogs also produced maps of western territories, the most comprehensive and famous of which was Capt. Gouverneur K. Warren's masterpiece, "The Trans-Mississippi West."

As Topogs were completing their work, Quartermaster General Thomas Jesup kept busy determining where to site frontier forts and their design, garrison, and resupply requirements. The continual expansion of settled territory made these activities difficult. Before the Mexican War, the U.S. Army manned and resupplied thirty-seven western posts, all relatively close to rivers navigable by steamboat. By 1855 the army occupied sixty-two posts in an area continental in size, with many posts in remote locations.[35] The use of eastern depots gave way to new army depots at Leavenworth, Santa Fe, and San Francisco. Normal resupply involved use of these depots for the storage of stock, local purchasing, and contracting for transport of supplies over vast distances using waterways, railways, and wagon trails to the western posts. The army spent millions annually on contracted civilian transportation, a practice that would continue during the Civil War. With the expansion of the army across the continent, the work of the Quartermaster Department grew to consume a third of the entire defense budget and contained the largest national bureaucracy. The War Department continued to second many line officers to the quartermaster general to construct forts, living quarters, and magazines and to build roads and bridges. These officers let contracts to hundreds of civilians in many remote military locations, sometimes being the largest employer in the area. Junior officers acting as assistant quartermasters, such as Lt. Ulysses Grant, were responsible for the logistics of a multitude of small garrisons that composed the military department of the West, and the skills these officers learned in performing such tasks subsequently shaped how they approached logistics problems in the Civil War.³⁶ Much of this work relied on subjects taught and skills acquired from the West Point curriculum, despite the fact that the military doctrine of this period did not address specific requirements of an army dispersed throughout the vast frontier.

No military doctrine manuals contained tactics for *la petite guerre* against western Indians. Army drill books for the infantry and artillery remained relevant to *la grande guerre*. Infantry doctrine still followed French designs. In 1835 General Scott had replaced the 1791 French infantry manual with his own version of the 1831 French publication. Artillery tactics were an amalgam of French and British horse artillery drills, compiled and refined in the 1840s by graduates Capt. Robert Anderson and 1st Lt. Miner Knowlton and by Maj. Samuel Ringgold. While not focused on *la petite guerre*, these works did at least offer standardization from which officers in western garrisons could make local deviations.³⁷ The secretary of war reinforced change in the mid-1850s when he commissioned Capt. William Hardee to produce a new infantry tactics manual for the rifled musket and improved munitions. Hardee based his *Rifle and Light Infantry Tactics* on the practices of the French *chasseurs à pied*, considered at the time to be the epitome of light infantry. The War Department convened a tactics review board at West Point in 1854 and made cadets demonstrate these drills. They approved Hardee's text as a supplement to Scott's *Infantry Tactics*, giving the army something of value for work in the West.

Professor Mahan had long understood that western service demanded independence of thought and action on the part of junior officers detached from their parent regiments. He proposed in the 1840s to augment the drill manuals with a book that gave regular and militia officers knowledge of small unit tactics applicable in any environment. Unable to secure War Department funding, in 1847 Mahan independently published his *Advanced-Guard, Out-Post, and Detachment Service of Troops*.³⁸ Eventually publishers printed four editions (8,000 to 10,000 copies), and Confederate printers pirated the text for use during the Civil War.

In writing this book Mahan drew upon his teaching notes. Given the practicality of the remainder of the text on *la petite guerre*, a reader might be surprised that he dedicated his first thirty pages to the historical evolution of conventional military thought from ancient times. Yet, during a period when the formal teaching of history was in its infancy, Mahan had the forethought to state: "It is in military history that we are to look for the source of all military science."[39] Here he illustrated the evolution of this science until its apogee under Napoleon, who had "fixed immovably those principles which . . . cannot fail to command success."[40] Napoleon had taken all that had existed before and perfected a military system that had become the basis of American military thought. But Mahan was emphatic that Americans could not emulate Napoleon's complete system because the emperor had no reservations about bloodshed whereas the United States, dependent on mobilized reserves, had to do what it could to preserve blood. Therefore, the task before Americans was to establish a system that would "do the greatest damage to our enemy with the least exposure to ourselves."[41]

Mahan acknowledged that an American military system existed, at least in part. It existed in Halleck's *Elements* and within his own publications on field and permanent fortifications, as well as in Scott's Americanized regulations and infantry drill manuals. What was lacking was a book about tactics for troops in detached service. He divided his work into chapters describing how to form and fight advanced guards, how to conduct reconnaissances, how to organize and escort convoys, and how to defend and attack isolated positions. Mahan desired with *Advanced-Guard* to teach junior officers to apply common sense and use latitude in decision making on these forms of service.[42] He presented his tactics as basic principles and not as detailed evolutions, formations, and drills. The key principle was the need for agility and celerity of movement by small groups of infantry, artillery, and cavalry with attached sappers and pontoniers.[43] Officers also needed to understand how to calculate weapons ranges, the widths and lengths of tactical formations, the lineation and progress of foot and horse movement, and how to achieve rapidity of fire. More than anything else, Mahan wanted regular officers and militia officers who were mobilized for operations in the rough terrain of North America to be able to apply good judgment. As such, the text was applicable to service upon the frontier.

Mahan did not write his *Advanced-Guard* specifically for detachments serving in the West. That would have been a shift away from doctrine for *la grande guerre*. Influenced by European theory and army practice in Mexico, his prescription for detachment work was within the context of a Napoleonic "grand corps" on offensive operations. Detachments were small bodies of troops working separately from their parent units in support of brigades or divisions moving to capture or secure a *key point*, hopefully upon the flank of an enemy army: "as an advance upon its front cannot be made without great loss and hazard."[44] The context of *Advanced-Guard* reflected a refinement in Mahan's thinking on the subject of "the science of war," now being called grand tactics and strategy interchangeably. In the 1850s, he desired to expand upon this and other subjects considerably.

Mahan attempted through all of his publications to disseminate doctrine on specific branches of military science to all elements of the peacetime military establishment. Up until 1840 he was content to keep his teaching and writing at what would now be considered the undergraduate level. Changes, however, came through both demand and personal interest before the Mexican War. In 1842 Totten had assigned responsibility to Mahan to provide advanced education to West Point engineer assistant instructors.[45] Totten, Delafield, and Mahan worked out a standard two-year curriculum requiring individual officers to meet Mahan biweekly to discuss readings and work out problems regarding the attack or defense of fortified sites. The readings were from Vauban, Cormontaigne, Bousmard, Poncelet, Pasley, Choumara, and Haillor—all engineers—and Archduke Charles and the campaigns of Napoleon and Wellington. Among other things, each officer undergoing this training had to design a complex permanent fortification for a given site and plan to attack it. This required extensive reading and recitation, including the latest theories of war.[46] The scope and content of these studies would today constitute graduate level work.

Mahan taxed his assistants. One claimed that "old Dennis set him to studying his eyes out on the principal battles of the world."[47] Thirty-seven engineer officers benefited from Mahan's graduate level tutelage in advanced military studies. It provided a more formal version of the instruction occurring simultaneously to a wider group in the academy's Napoleon Club. Of the thirty-seven engineer officers under Mahan between 1840 and 1861, twenty-five rose to the rank of brigadier or major general during the Civil War (including seven corps commanders and three army commanders; see table 7), with the remainder serving in senior engineer and staff positions during the war.[48]

Mahan also sought ways to make knowledge of military science accessible to the cadets and to interested militia and volunteer officers. Therefore, during the last decade of the antebellum period, his thinking regarding the instruction of permanent fortifications evolved. He submitted to the reality that, in terms of formal military education in United States, his course for cadets would probably constitute the sole conduit for instruction in permanent fortifications available. Therefore, he stopped teaching multiple systems that would only serve to confuse his students. He simplified fortifications instruction from how European schools taught it, concentrating on fundamentals, which he encouraged cadets to apply in any situation, allowing adjustments to meet local conditions. He abhorred tendencies to "beget a servile spirit of method" and instead solicited a "judicious adaptation of these principles and elements."[49]

Between 1830 and 1835, Mahan used his lithograph notes to augment Gay de Vernon. He then published his field fortifications manual, and this along with his lithographs became the course texts. In 1850 he published his notes on permanent fortifications, making them accessible to everyone. In 1863 he published a two-volume edition of military engineering that reiterated both field and permanent fortifications and added an addendum on strategy and grand tactics to a new edition of *Advanced-Guard*.[50] These notes and texts were sufficient for the teaching of cadets and for reading by reserve officers. For the engineer officers studying under him, Mahan used more advanced works available in the academy library.[51]

Mahan demanded that cadets learn first principles by rote, but once mastered he was not opposed to dialectic. For Third System purposes, where local conditions would require variation of fortifications design, he was acutely aware of need for debate, innovation, and improvement. However, he was suspicious of motives for debate, implying that much of it was self-serving to engineers who advocated reforms simply to have their names enshrined in a new fortifications method.[52] Mahan was at the same time aware of the possibility that debate regarding innovation might, if introduced too early in one's education, come at the expense of sound knowledge and confidence in first principles. Dialectic that required synthesis by the individual could lead to variety of interpretation—contributing to weakened standards, no common base knowledge, and the forsaking of principles. Here historians can see something more than conservatism but also the Enlightenment need for knowledge that is applicable and not so abstract as to prevent application. In the postmodernism of our times, it is difficult to appreciate this worldview. Yet it is exactly this idea of standardized mental outlook that allowed military doctrine to flourish in the nineteenth and much of the twentieth century. That some western military doctrines became dogmatic (the German pursuit of battles of annihilation and the French *attaque à outrance*, for instance) does not diminish the important unifying effect of doctrine, and Mahan remained conscious of such purpose.

Heavily committed to the enhancement of engineering theory and practice and the instruction in permanent and field fortifications, Mahan also supported experimentation in artillery and small arms ballistics and the advance application of fire, particularly as they affected fortifications. Experimentation was occurring throughout the decade. It is interesting that while Antoine Jomini was reluctant at this time to accept that advances in weapons fire were changing the nature of combat, the more scientifically inclined Mahan was stating outright before the Civil War that open assault tactics of Napoleonic battle against field fortifications would produce slaughter because of the longer range of small arms.[53] His beliefs led him to reemphasize the role of field entrenchments in combat, even for the army on the offensive.

The Teaching of Strategy at West Point

Throughout the 1850s Mahan refined his thinking about military strategy and the indivisible field of logistics. Staying true to the work of Bülow and Archduke Charles, and a little distant from the work of Jomini, Mahan had reduced European complex systems regarding lines and bases of operation. He distilled everything down to a simple concept regarding the relationship between the base and lines. The United States, as with all European countries, would fight her wars along or from frontiers, whether maritime or inland. Preparation for war necessitated the militarization of a particular frontier into one or more bases of operation—one or more "strategic points" from which an army could launch an offensive campaign by moving from the base to an "objective point" that provided a position of advantage threatening the enemy's line of communications. Giving battle was a part of such campaign design, but contemporary thinking believed that decisive battle would come only when an enemy's army was unable to fall back on its base of operations.

Mahan's teaching encouraged the turning movement gained by strategic marches, planned using the geometry of lines projected from bases. A key feature in Mahan's later thinking on strategy was his flexibility about the use of alternate bases of operation lying oblique in relation to the primary base and upon the flank of the enemy's line and base. The capture of Vera Cruz in Mexico had achieved just such an alternate base in relation to Buena Vista and Santa Anna's army and Mexico City. An alternate base allowed an advancing army the option to sever itself from a primary base and to maneuver to link up with a new base, giving that army greater flexibility to turn an enemy. This concept had indeed been part of Jomini's *Summary*, but was obfuscated by Jomini's abstract machinations regarding the advantages of the central position and interior versus exterior lines. In his appreciation for proper bases and alternate bases, Mahan was closely associated with French general staff officer Auguste Frédéric Lendy. Lendy's *Principles of War; or, Elementary Treatise on the Higher Tactics and Strategy (1853)* gave superb and clear descriptions of the value of well-selected bases of operation and the advantages wrought by an alternate base perpendicular to that of the enemy. Use of alternative bases of operation constituted a fundamental shift away from the sanctity of Jomini's interior lines. Generals Grant and Sherman would later demonstrate mastery of its use.[54]

Mahan's simpler rendering of lines and bases was not unique. It was similar to that being advocated by G. H. Dufour in his *Cours de Tactique* (1841), which was in use in Europe and available at the West Point library. Dufour's work was so acceptable to Mahan that when looking for a précise to outline strategic theory during the Civil War, Mahan translated it and added it as a supplement to his 1863 edition of *Advanced-Guard*. This he did without accreditation.

Mahan's insistence on an integral relationship between the base and the line of operations was an extension of his intimate understanding of—and faith in—Vauban's siege methods, claiming that "those reciprocal relations are very much the same as those of the parallels of a siege to the approaches on the place."[55] He believed that the proper identification and use of the base of operation theory was of fundamental importance to the successful prosecution of war, even in *la petite guerre*. He felt, for instance, that the preferred strategy for dealing with the Native Indian was to strike into the heart of their territory and dismantle their base, denying them any means of subsistence, and forcing their surrender.[56]

Mahan's emphasis on logistics (the fundamental purpose of the

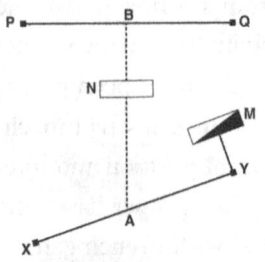
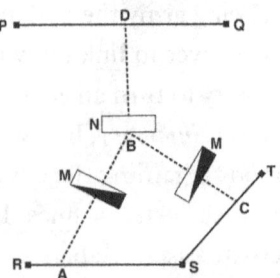
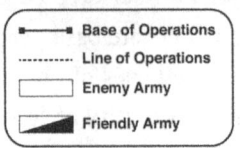

FIGURE 6. The evolution of Mahan's ideas of strategy illustrating the advantage of the oblique and concave (or secondary) base of operations. Based on "Principles of Strategy and Grand Tactics" by Dennis Hart Mahan in *Advanced-Guard, Out-Post, and Detachment Service of Troops: With the Essential Principles of Strategy and Grand Tactics for use of the Officers of the Militia and Volunteers* (New York: John Wiley, 1863), plate 7, figs. 15 and 16. Drawn from the English edition of G. H. Dufour's *Strategy and Tactics* (New York: D. van Norstrand, 1864), originally *Tactique (Cours de)* (Paris, 1841).

base of operations) reflected both his modernity and the reality of vast American frontiers. No matter which branch they served, American officers had to think about logistical planning before, during, and after all operations. The theory of the science of military movements aided their thinking by providing a simple yet useful guide, applicable to the United States.

The theory of the base of operations was an important part of Mahan's thinking and teaching. It would leave a mark on subsequent American military thoughts of war. However, as in all things, Mahan suggested common sense and acknowledged the power of the sublime in the application of the theory:

It is in military history that we are to look for the source of all military science. In it we shall find those exemplifications of failure and success by which alone the truth and value of the rules of strategy can be tested. Geometrical diagrams may assist in fixing the attention, and aiding by the eye the reasoning faculties; but experience alone can fully satisfy the judgment as to the correctness of its decisions, in problems of so mixed a character, into which so many heterogeneous elements enter.[57]

The academy's five-year program of study commencing in 1854 allowed the faculty to expand the teaching of strategy. The program introduced new aspects to the West Point curriculum, namely, Spanish and limited recitation in English studies, moral and intellectual science (history of philosophy), and law. Extra time was allocated to practical exercises in cavalry, infantry, and artillery tactics, practical engineering, and ordnance studies (see table 3). Civil engineering was kept within the fourth year of study but was reduced in scope as civilian colleges began to teach the same topics. In the fifth year, cadets had a three-hour period each day for the entire year on military engineering and the science of war. Separating civil engineering from his teaching to senior year cadets, and shifting practical military engineering to another department, allowed Mahan the opportunity to introduce advanced ideas of strategy in the fifth year. In this, he used lectures and maps painted on walls of recitation rooms to help illustrate historical examples.[58] The lectures covered past battles and campaigns and highlighted maxims and principles such as the theory of the base and lines of operations. Records show that these lectures eventually included the battles of Leuctra, Zama, Pharsalus, Morgarten, Fontenoy, Lethen, Bleheim, Arcola, Austerlitz, Friedland, Wagram, Waterloo, Gettysburg, Vicksburg, Nashville, and Sadowa.[59] During the five-year program, from 1854 to 1861, and during the Civil War, cadets began to use military history in their studies, reinforcing principles and lessons from past campaigns. The academic papers of Charles Dempsey (class of 1865), for example, each twenty to thirty pages in length, presented such topics as the defense of Charleston Harbor, the British incursion into Spain, 1808–9, the second French siege of Saragossa, 1808–9, and the invasion of Spain by Napoleon, 1808.[60]

The West Point Doctrine Compared

West Point was not the only place where discussions on strategy were occurring. Components of civil society in the United States had long been interested in military matters, starting at least with subscriptions to the *Military and Naval Magazine of the United States* of 1833, which was absorbed into the *Army and Navy Chronicle* in 1834 and replaced in turn by the *Army and Navy Chronicle and Scientific Repository* in 1843. Thereafter American readers could acquire European military, philosophical, and technical periodicals directly or read articles in standard American journals such as *Harper's Weekly*. Government was also interested in military developments. Jefferson Davis, himself an academy graduate and hero of the Mexican War, was appreciative of military science and wanted to acquire the latest European thinking on a wide array of technical subjects. In 1854 he selected three officers to go overseas, each representing a branch of army service. A former superintendent of the military academy, Maj. Richard Delafield, led the commission and reported on matters of fortifications and engineering. Artillery Maj. Alfred Mordecai reported on ordnance and artillery issues, and Capt. George McClellan, newly transferred from the engineers to the cavalry, reported on developments of European infantry and cavalry. The commission was given liberty to travel extensively, including a visit to the Crimea, to the scene of the last great European war. Each of the three produced separate reports.

The report of Major Mordecai was very technical, meeting all of Davis's requirements. He concentrated on the design and manufacture of weapons, the maintenance of national depots, and the latest developments in small arms and artillery weapons and munitions. He was thorough yet cautious regarding the use of rifled muskets, seeing promise in their development, but suggesting that experimentation was required before the army could make recommendations regarding their use by large, rudimentarily trained volunteer armies. He did note, however, that their use during the siege of Sebastopol demanded extensive employment of earthen field works, "rifle pits," and trenches to protect soldiers from the effects of longer-range rifle fire.[61]

Perhaps the greatest result of Mordecai's report was its affirmation of the efficacy of a new French 12 pound gun-howitzer that was replacing the 6 pound guns and older 12 pound howitzers in the U.S. Army, easing ordnance and logistics problems and giving the army more agility and capability. This weapon, known as the "Napoleon," became the favorite artillery piece of the Civil War.[62]

McClellan opened his report with a recounting of the siege of Sebastopol, which he believed to be the whole objective of the Crimean War (everything before that being preliminary and setting conditions for this climax) and foreshadowed his thinking in the 1862 Peninsula Campaign. McClellan then treated infantry and cavalry organizational, technical, logistical, and tactical innovations of the Russian, Prussian, Austrian, and French armies and the English cavalry. He also dwelt on the construction of specialized barracks and equipment. He observed that in comparison with the excellent disembarkation and logistics build-up for the siege at Vera Cruz in 1847, the British and French landings in Crimea and their logistics systems were poor indeed. Remarkable in his report is the lack of mention of field fortifications outside of his historical analysis of Crimea, where he lavishly praised Russian use of earthen defensive works. He left engineering considerations to Richard Delafield.

McClellan drew two major lessons from the Crimean War. The first was that a system of permanent fortifications was "wise and proper." The second was the importance of rigorous military education, because "mere individual courage cannot suffice to overcome the forces that would be brought against us, were we involved in a European war." McClellan believed that command in war "must be rendered manageable by discipline, and directed by that consummate and mechanical military skill which can only be acquired by a course of education, instituted for a special purpose, and by long habit."[63]

Major Delafield's report was also concerned with being prepared for *la grande guerre* against a European power, highlighting America's "comparative want of preparation and military knowledge." Delafield believed that such a war was inevitable and that "the Secretary of War will do a great service to the nation by increasing the material and munitions, means of defence, and the diffusion of military information in every possible way . . . without creating any more of a standing army than the growth of the country calls for."[64] Focusing on fortifications, Delafield showed just how seriously Europeans were about the improvements of permanent works, something he felt the United States needed to emulate. Delafield gave a detailed report on the latest innovations in military architecture and equipment in Europe, including plans for various types of barracks, hospitals, ambulance construction, and the latest mechanical and technological innovations for wagon and boat design. He noted no new developments in the general art of war since Napoleon, only an increase in the "magnitude of the engines of war."[65]

The Delafield Commission Reports were published just in time to rebut the latest round of attacks on West Point and Mahan, this time by one of Mahan's graduate students, Capt. James St. Clair Morton. Morton had a benefactor in Secretary of War John B. Floyd (1857–60) and was able to put a critique of West Point instruction into the 1857 Board of Visitors' Report, and with Floyd's support, he was able to publish several critiques of fortifications theory in the United States. Specifically, he challenged the continuance of teaching continental European methods, stating his preference for the British vocational approach to engineering, based on experience more than theory. He was also an advocate of the steamship, floating steam-powered batteries, and earthen defenses rather than permanent coastal fortifications. His published critiques led to a court of inquiry, presided over by Col. Robert E. Lee, which exonerated Mahan and the Fortifications Board's Third System policy. One testimonial, given by Maj. George McClellan, freshly back from his visit to Crimea, claimed that Mahan's course was better than any courses conducted in Europe.[66] McClellan's faith in Third System thinking was firm: "The permanent defences of the harbor of Sebastopol against an attack by water . . . seem to establish, beyond controversy, the soundness of the view so long entertained by all military men, that well constructed fortifications must always prove more than a match for the strongest fleets."[67]

Morton's public charges probably caused Mahan to be more conservative at a time when he was prepared to modify his teachings. He was in the process of revising his 1850 text on permanent fortifications to include lessons from Europe, especially Crimea. In the revised text—prepared in the late 1850s but published during the Civil War—Mahan acknowledged the vulnerability of masonry to new shell technology, and this reinforced his emphasis on hiding masonry behind dirt ramparts. He was irritated when others claimed that a key lesson of Crimea was that earth was the best protector against artillery. In an article for *Army and Navy Journal*, he stated that it was not Crimea that enlightened all to earthen works but "a well-settled maxim of engineering, from the time of Vauban, to cover all masonry by earthen masks."[68]

Apart from challenges to fortifications theory, sectionalism threatened West Point. In 1857 the legislature of Tennessee proposed the establishment of a western branch of the U.S. Military Academy exclusively for cavalry and infantry cadets.[69] In response, Richard Delafield submitted to the Committee on Military Affairs that a second western academy would only serve to promote sectionalism. Delafield argued that it would be dangerous to the all-arms focus of army education to have separate arms academies. In this, he referred to his official visit to Europe by comparing their schools with West Point:

Its [West Point's] striking characteristic is the education of cadets for *all arms* in contradistinction to special schools for each arm. By our system the cavalry officer in the field, finding himself in command of artillery and infantry, by the chances of war may lose his artillery officers, when any graduate of this institution can at once command the battery. And so with any combinations. Each and every officer knows enough of the theory and practice of other arms to command or fill subordinate stations under any contingencies. Infantry, cavalry, or artillery graduates can each resort, in cases of necessity, to the engineer and ordnance officers' duties—throw up intrenchments, and prepare ammunition or fortify himself; while, at the same time, each or any of them knows the strength of such works as they may have to attack. Our present institution is . . . superior to anything that can be found in Europe.[70]

Delafield's comments were echoed a year later in the final review of West Point military science before the Civil War. In 1860 an investigative commission led by Jefferson Davis endorsed the West Point method, but suggested improvement by ceasing the five-year program in favor of the old four-year model. The review included a comparison of the academy with European institutions. Specifically it compared the curriculum with the two-year French school at St. Cyr (for infantry and cavalry officers), the two-year school at Metz for French artillery and engineer officers, the school at Woolwich for English engineers, and Prussian, Austrian, and Russian schools for young cadets. While this comparison showed that West Point was not unique in its emphasis on military science—with most schools requiring some advanced mathematics, topography and geography, physical sciences, fortifications, engineering, military history, administration, and law—it remained the only one demanding that all cadets learn all-arms tactics. This comprehensive review found no great deficiencies in the West Point program of study, but recommended an increase in the teaching of strategy.

Specifically, it reiterated the Enlightenment view of war by recommending that "no material reduction of subject or time can be made in the mathematical and scientific studies and their application to the art of war, if we are to keep pace with the advance of science."[71]

The Davis Commission confirmed that West Point was providing the basics that a professional officer required for the performance of duties in any branch of service. However, the commission report did repeat an observation made frequently in the antebellum period that the army also required institutions for advanced military learning, in the form of branch schools. The infantry, artillery, and cavalry required "schools of practice" along the European models, each dedicated to taking basic skill and knowledge provided by West Point and expanding these in more advanced battlefield applications of a specific arm periodically throughout a career. Secretary of War Calhoun had established just such a school for artillery in 1824 at Fort Monroe, and an infantry school opened in 1825 at Jefferson Barracks, St. Louis. Like so many others to follow, however, these schools were experiments and failed to receive governmental support once administrations changed. Fort Jefferson ceased to function as an advanced school in 1829 and Fort Monroe in 1834. The cavalry established a Mounted School of Instruction at Carlisle Barracks, Pennsylvania, in 1838, but this too suffered from want of regulation demanding specific attendance or standards. Congress and the War Department might agree in principle with the requirements for advanced schools of application, but no funding and legislation were forthcoming. Throughout the antebellum era, West Point provided the singular institution of continuity.

By 1860 the military science taught at West Point was becoming more widely disseminated throughout the army and into the volunteer militia organizations. Mahan's treatises, as well as Hardee's, Halleck's, and Anderson's works and Scott's *Infantry Tactics*, were the standard texts of the regular army and state forces. Napoleonic in character, containing a great deal of French tactical method, they were also Americanized, written to be applied to war in North America. That said, they retained their focus on *la grande guerre* and linear tactics, the epitome of warfare being defensive and offensive lines that could deliver effective small arms and artillery fire. When taken together, the texts covered both tactical and strategic prescriptions and outlined the basis of strategy, at least campaign planning and movements. Beyond military treatises, military science was manifest in the work of the topographical engineers in western exploration and railway surveys and in the hundreds of fortifications and internal improvement projects that formed the frontiers of the United States. Connected with this was the considerable effort that the U.S. Army was putting into ordnance development through experimentation with new calibers of weapons and new material for the casting of cannon. Always with the aim of mobilization in mind, the army continued to produce thousands of weapons every year in federal foundries. Even with the closure of the Harpers Ferry weapons plant after the John Brown incident, the Springfield plant produced 14,615 new rifled muskets in 1860. In the fifteen arsenals and depots administered by federal forces in 1860, there were more than 530,000 stands of arms to distribute to mobilized state forces if war came (135,000 of these in Southern states).

The responsibilities of the regular army officer corps in the event of mobilization were well known. The War Department expected even relatively junior ranking officers to organize volunteer troops into battalions and to provide each with regulation. It also expected army officers of every branch of service to act as quartermasters, commissariat, paymasters, and assistant adjutants general when required. The problems of mustering volunteers and providing uniforms, weapons, training, discipline, transportation, encampments, and supply in the field were the responsibility of regulars. Luckily, many of these officers had practice in these very tasks as they circulated through general staff appointments in between line assignments throughout the antebellum era. Over 43 percent (850) of the 1866 graduates who had served between 1802 and January 1861 (and a much higher percentage of the long-service officers) had considerable prewar experience as instructors at the academy, as officers performing engineering or topographical engineering duties, or as officers performing important logistics, ordnance, and administration staff functions—all preparatory experience for the challenges of national mobilization.[72]

By 1860 the public and their representatives in Washington considered West Point a valuable national institution despite sectional tensions that created a desire for stronger state military establishments and despite debates regarding priorities of ships and coastal fortifications and the validity of Vaubanian constructions. Military science dominated professional military thinking during this period and remained unchallenged as the basis for military instruction. The academy faculty made modifications to the curriculum to include advanced instruction for assistant professors and a new five-year curriculum that allowed subjects other than science (Spanish, English, logic, and military history) to be included. Contemporary publications by academy graduates also captured the state of American military science. While differences of opinion continued over details, public review of the academy's curriculum confirmed its utility.

On the verge of the Civil War, the four distinct categories of military science that had emerged out of the Military Enlightenment in France in the eighteenth century had been successfully transplanted in the United States. They had grown roots and sprouted an American tree of knowledge at West Point. Graduates of the academy, inculcated in a doctrine of military science, served throughout the nation performing duties that had them apply one or more of its aspects, so much so that military science became their paradigm of war. The following chapter will suggest how this experience was influential. It will show that the work of these officers, including their substantial writings, ensured that throughout the

U.S. Army, and in many states' militias, there existed in 1861 considerable expertise in fortifications and engineering, artillery applications, the logistics of mobilization, and the science of strategic movements. The fruits of this tree of military science were about to be tasted by millions of Americans.

8

Military Science in the Civil War

One can only truly understand the Civil War through an appreciation of military science. I do not suggest a particular decisive effect or that military science determined the war's outcome, only that the West Point graduates serving in key positions on both sides used the doctrine they had learned at West Point in conducting operations from the beginning to the end of the fighting. Demonstrating this for every operation of the Civil War would be a massive undertaking. Not beyond the depth of my research, such an effort is however beyond the scope of this book. Therefore, it must suffice here merely to lay down markers that encourage future analysis, research, and writing.

West Pointers in both blue and gray used elements of military science throughout the entirety of the war. They followed mobilization procedures laid down in Third System policy and in antebellum tradition, and sustained war efforts using conscription, advances in ordnance, and the militarization of industry. They drew from contemporary theories of strategy to plan campaigns by identifying strategic points, establishing bases of operation upon expedient military frontiers, and making lines of operation follow improved rail, road, and waterways. West Pointers utilized all the modern advances in transportation and communications technology to support military logistics. They used permanent fortifications extensively—especially upon frontiers and as bases—and conducted sieges in the manner of Vauban. At the tactical level, West Pointers taught volunteer armies how to use expedient fortifications and field engineering to both arrest and facilitate the movement of forces. Graduates evolved army organization toward the Napoleonic all-arms corps model and created the first field armies. Throughout the war, West Point generals sought to implement Mahan's prescriptions of celerity of movement, maintenance of the initiative, and concentration of force.

Yet the application of military science was no guarantor of success. As emphasized in prewar manuals, learned knowledge only served to enable sublime personal talents. Many Civil War commanders, regular army and political appointee alike, lacked the character traits necessary for high command. Many West Pointers, well schooled in military science, let personal ambition replace good generalship, or succumbed to weakness when faced with the reality of war, becoming paralyzed when required to make decisions in the face of the enemy, or they simply lacked resilience. Many others proved to personify Halleck's *practical men* and *martinets*. The most successful Civil War generals were those that demonstrated proper qualities of decision making, courage, and determination, and for a number of reasons, most were West Point graduates. While some historians steadfastly oppose the notion that these commanders were the least bit affected by antebellum doctrine, the conduct of operations between 1861 and 1865 shows that the paradigm of war shared by regular officers in 1860 endured and at least somewhat shaped military events from Sumter to Appomattox. The West Pointers who rose to command, and the hundreds of officers who supported them in staff positions, had each earned the equivalency of a basic engineering degree, and many had considerable staff experience or instructional experience in subjects related to the science of war. All were comfortable with advances in technology and most embraced them. At West Point, these graduates had learned that a professional soldier had a duty to prepare for war, and that as professionals they needed to stay aware of how advances in science were steadily changing the art of war. This Enlightenment faith did not wane in the war years. Therefore, while contingent forces still held sway over how West Pointers performed, it is simply wrong to deny the influence of their shared understanding of war.

West Point's Role in the War

Between April 1861 and May 1865, 1,135 graduates served either the North or the South in some capacity.[1] Of these, approximately one-third (374) had prewar experience on military operations, and over a quarter (306) had considerable prewar staff, engineer, or instructor experience.

Between February and April 1865, Congress awarded a plethora of brevet ranks to Union officers as reward for good service. Before that distorting event, almost one-third (368) of the 1,135 graduates who had participated in the war had become general officers, commanding brigades, divisions, corps, and field armies. Four became commanding generals. In addition, West Pointers commanded most geographic military departments and districts. The majority (265) of these 368 commanders had prewar experience above the drudgery of company level management; 129 had served in staff department positions; 105 had experience in engineer or Topog duties; and 94 had been instructors at the academy (see table 8). This group of 265 included the most successful generals of the war. Almost all had participated in Professor Mahan's course on the science of war. Many had participated in his advanced engineering class or in the Napoleon Club (see tables 3 and 4) and came to the war with graduate level understanding of military science. They commenced the Civil War with something more than an appreciation for company and battalion tactics and approached operations with a specific paradigm. However, they did not draw solely from what West Point had instructed. Their antebellum staff and line service in a variety of settings had also taught them to deal with real engineering, logistics, fortifications, and general staff problems.

MILITARY SCIENCE IN THE CIVIL WAR

Historians have cataloged the rise to command (and hard falls) of many of these generals. Less well known are the remainder of the 1,135 graduates who served in the war. Of these, 797 served as staff officers; 92 served with the general staff departments in Washington or Richmond; 155 served on geographic departmental staffs; 211 on district staffs; 264 were field army staff officers; 23 were army chiefs of staff; another 57 commanded depots and arsenals. Of these 797, 46 percent (371) had prewar staff experience; 126 having served as academy instructors, 132 had performed engineer duties, and 149 had worked for the staff bureaus. If it is an unfortunate characteristic that history judges commanders against a hard list of personality traits, with few measuring up. It is equally unfortunate that staff officers are unmeasured, underappreciated, and even largely ignored.

West Pointers were well suited for such staff appointments. Their reputation for honesty and accountability, skill in mathematics, and their common understanding of all arms of the army gave them advantage over state-appointed officers. They shared a military language and an understanding of standard processes. The commonality of thinking about *la grande guerre* amongst West Pointers formed the basis for large-scale staff planning that gave the Civil War something more than what the vagaries of individual personality would have provided. West Pointers functioned in every sense as a modern general staff, sharing a mindset about how to raise, organize, train, supply, move, encamp, feed, and fight armies of volunteers and conscripts—all of the elements of military science they had learned at West Point.

Mobilizing and Organizing for War

The Civil War began with an artillery duel over Third System fortifications in Charleston Harbor in one of a number of standoffs and exchanges over the seizure of federal military infrastructure by seceding states. The war ended after a great siege around Petersburg-Richmond, an engagement that would have been familiar to Vauban. Between these events armies formed, marched, and fought a protracted war of a scale that very few foresaw. Initial mobilization of Southern and Northern armies followed the principles long-used in America, at least until conscription was invoked in 1862 in the South and in 1863 in the North. Comparatively few of the 16,000 regular soldiers joined the Confederacy, preferring to serve the Union, and in April 1861 Lincoln initially called for 75,000 three-month volunteers to augment these troops, asking for 400,000 more that July. Jefferson Davis also called for hundreds of thousands of soldiers, and subsequent calls for volunteers and conscription on both sides eventually put millions of soldiers in uniform. With each early call, governors organized the muster of state volunteers at central locations where federal authorities could take control. The governors (and federal politicians) also appointed general officers to command the volunteer forces, and during this first year, it was common practice that volunteer soldiers elect regimental officers.

The scale of these call-ups was unprecedented and supplying these forces quickly exhausted federal stockpiles, forcing governors and quartermasters of militia to tender contracts to private businesses within the state, or to employ agents to seek contracts in other states or abroad, to procure equipment for their state volunteers. Lincoln's Indemnification Act of 27 July 1861 encouraged individual Northern states to provide such supplies with the promise of federal reimbursement. This decentralized system had worked well enough in raising forces for the Indian wars of the 1830s and 1840s, the Mexican War, and the military expeditions and campaigns of the 1850s. War departments North and South readily accepted the role of states in the generation of forces. However, as mobilization grew throughout 1861, considerable contracting abuse followed. Politicians with fists full of cash were eager to parcel out favors, and entrepreneurs were equally eager to secure lucrative contracts while keeping production costs low, leading to the manufacture of much substandard equipment at inflated prices. Certain that the war would be over in a matter of months, governments stomached such things. Southern states were particularly resistant to change these decentralized procurement methods, with North Carolina refusing to accept any other mechanism to supply her regiments. Insistent upon the perpetuation of states' rights theory into war making, and almost totally reliant upon cottage-industry production and foreign purchase, logistics in the Confederacy remained inefficient compared to that of the Union.

In late 1861, Northern leaders recognized the problems related to decentralized procurement and the War Department issued orders that the states handover mobilization to U.S. Army officers working for the quartermaster general (Maj. Gen. Montgomery Meigs) and other bureaus in Washington. Between autumn 1861 and spring 1862, army officers assumed all mustering, disbursing, inspection, commissariat, pay, and depot and arsenal management duties. The higher ranking of these officers were West Point graduates. Of the 1,135 spoken of above, the Union employed 94 as district, state, and department quartermasters and commissariat officers, 31 commanded depots and arsenals, 64 worked in ordnance functions, and 93 became superintendents and assistant inspector generals within the geographic departments.[2] These officers took control of procurement contracting, enforcing uniform standards and specifications of production, and overseeing a mix of contracts to private business and to federal-funded businesses that made items directly for the War Department. Responsible to Congress through Meigs and other bureau chiefs in Washington, these officers eliminated most of the abuses and corruption that previously existed.

The national logistics system that emerged from these reforms militarized the Union's economy in a manner never before experienced or contemplated. It gave enormous powers to the federal government. The quartermaster general became the single biggest employer in the Union, paying more than 100,000 employees to make clothing, equipment and wagons, to purchase horses and mules, and to move large quantities of men and material throughout the North.[3] The quartermaster general's work was but one part of a large system that allowed for rapid expansion of war-related industry. A boom in the production of oil and iron ore as well as the availability of factory labor and the ease of moving commodities on an expanding railroad network supported an increase in manufacturing. By mid-1862, the quartermaster general and the head of the ordnance bureau had harnessed the capacity of Northern industry, ensuring that Union soldiers never wanted for materiel and food (with the singular exception of the "Cracker Line" to besieged Chattanooga in autumn 1863). That year Meigs was able to stockpile 300,000 uniforms in three central depots (St. Louis, Cincinnati, and New York), ready to ship to necessary mustering points. Federally owned and contracted firms working for the ordnance bureau achieved staggering annual production rates of weapons and ammunition. In 1863, 1,082,841 rifles and muskets and 1,577 cannon were manufactured, and 254 million cartridges and 1.5 million cannonballs were made.[4] During this period of centralizing reform, Union Army mustering and disbursing officers (53 of them West Pointers) assumed responsibility for recruiting in the states under governor oversight with coordination by the adjutant general's office in Washington. More than 2.1 million Union soldiers were enrolled by April 1865.

Antebellum notions regarding the importance of war preparations, as well as the practices of logistical sustainment and transportation of forces between numerous outposts spread across the vastness of the West, gave Union army officers knowledge about how to expand the states' mobilization mechanisms and combine them into a more potent centralized system. Instead of mustering forces for movement to threatened coastlines, these officers enrolled, gathered, equipped and transported these volunteer units to bases upon the frontiers forming between the Union and Confederacy. If the epitome of war was the concentration of mass upon key strategic points, the creation of the mass was being achieved by the largely unsung work of officers flung throughout the states in jobs as important as they were lacking in glory. They recruited and mustered the soldiers, tendered and monitored the equipment contracts, and ensured quality control in a process that made soldiers out of cit-

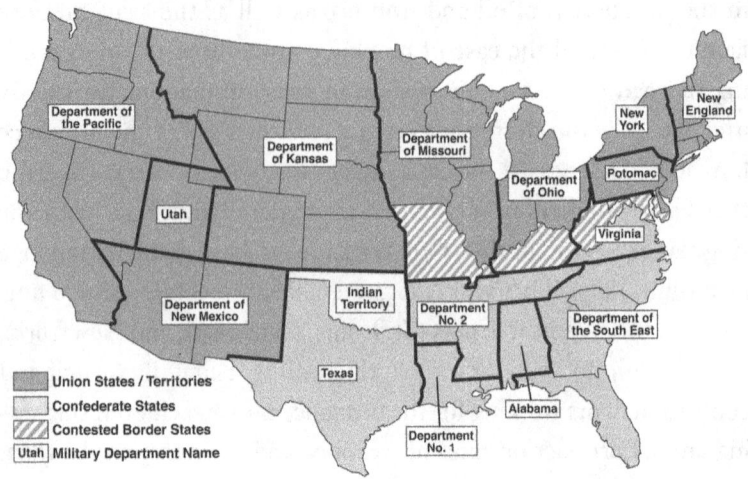

MAP 7. Military departments, December 1861. By 1865 the Union had redesignated the eastern theater as East, Middle, Washington, Virginia, North Carolina, and South departments, and the western theater of war contained the Northern, Kentucky, Mississippi, Gulf, Arkansas, Missouri, New Mexico, and Pacific departments. Based on *The Official Military Atlas of the Civil War* by George B. Davis et al. (New York: Gramercy, 1983), plate 164.

izens. Their efforts made it possible for the generals commanding armies, and their political masters, to think about how to employ these masses using strategy.

Organizational and topographical factors were shaping Union and Confederate strategic thinking. Both sides grouped states together into regional departments, each under the command of a senior officer responsible for the coordination and management of all war efforts of the states within his region. The department names and boundaries changed as the war went on. The Union started with departments in New England, New York, Potomac, Ohio, Missouri, Kansas, Utah, New Mexico, and the Pacific (see map 7). The departments of the Confederacy included North Virginia, North Carolina, South Carolina/Georgia/Florida, Alabama, Department No. 1 (Louisiana), No. 2 (Mississippi, Arkansas, and Tennessee), Texas, and the Indian Territory (Oklahoma).

While convenient for helping to organize state-run mustering and supply, these departments were not optimum for formulating strategy, planning campaigns, or conducting operations. Eventually military activity coalesced into what historians have referred to as the eastern and western theaters of war. Eventually, too, departmental forces amalgamated into large field commands aligned to the two theaters. By 1864 two grand Union armies emerged as the means of defeat of the Confederacy. In the eastern theater, the Army of the Potomac grew to consist of seven corps comprising 150,000 men. In the west, the Army of the Tennessee combined with the Army of the Cumberland and the Army of the Ohio to give the commanding general, William T. Sherman, more than 100,000 men. Other smaller subsidiary military commands existed in each theater; all, however, in this year came under the supreme command of the general in chief, Ulysses S. Grant. Confederate field armies operating in the western theater (less Texas and Arkansas) remained largely independent, with the principal force being the Army of Tennessee. In the east, the principal Confederate military force was the Army of Northern Virginia under Robert E. Lee.

The demarcation of departments, theaters of war, and field armies had less to do with strategic science than with convenience of organization and the undeniable reality of topography. The physical barrier of the Allegheny Mountains divided theaters, and coastlines and rivers added further natural boundaries. The divides helped to create frontiers. The Confederacy held maritime frontiers along the east and gulf coasts, where Third System fortifications found great relevance. Another frontier in the east ran initially along the Potomac River to West Virginia. In the west, the land frontier ran, initially, along the Ohio River and northern Tennessee border (Kentucky being neutral in 1861) to the northern Mississippi then along the northern Arkansas and Texas borders.

Both sides established depots from which to transport forces and equipment to these frontiers. The Union built depots (where much of the work of department quartermasters, ordnance, commissariat,

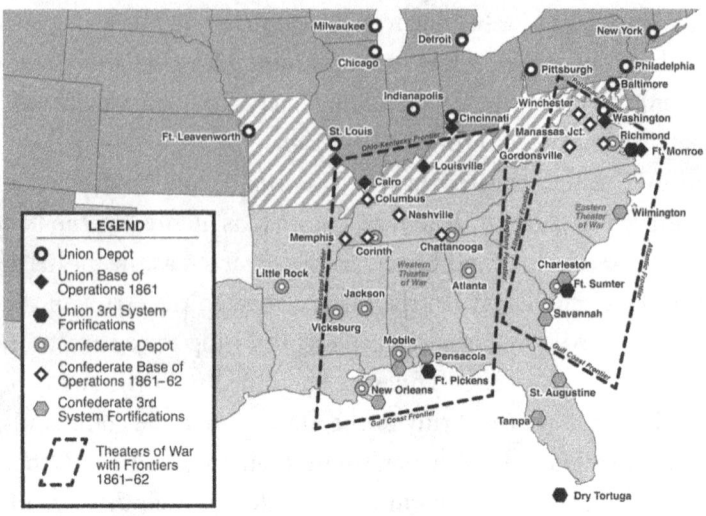

MAP 8. Frontiers, depots, bases of operation, and Third System works, 1861–1863. Based on *The Official Military Atlas of the Civil War* by George B. Davis et al. (New York: Gramercy, 1983), plate 164.

and mustering officers came together) at New York, Philadelphia, Baltimore, Washington, Pittsburgh, Cincinnati, Detroit, Indianapolis, Chicago, Milwaukee, and St. Louis. From these depots, soldiers and material fed Union field armies forming at bases of operations at Washington, Cincinnati, Louisville, and St. Louis (then Cairo). The Confederates maintained depots at Richmond, Charleston, and Savannah in the east and at Atlanta, Chattanooga, Mobile, New Orleans, Jackson, Corinth, Vicksburg, and Little Rock in the west, pushing troops and material to forward bases of supply at Manassas Junction, Winchester, and Gordonsville in Virginia (with Richmond remaining as the principal base of operations), and Nashville, Corinth, Memphis, and Columbus in the west.

Antebellum military theory emphasized the importance of fortifying bases of operation (alternately referred to during the war as "bases of supply") and connecting bases and depots using communications networks. Fortified and connected bases served both defensive and offensive purposes, they defended against enemy advances, and they allowed provisions to be stockpiled before and during offensive campaigns. Naturally, Third Systems fortifications were suitable as bases of operation, and the Union made both defensive and offensive use of Fortress Monroe at Hampton Roads before and during McClellan's 1862 Peninsula Campaign. The Confederates relied heavily on fortifications on the east and gulf coasts—especially permanent works at Wilmington, Charleston, Savannah, Tampa, Pensacola, Mobile, New Orleans, and Galveston—to prevent Union seizure of strategic points and to protect the Confederate Heartland. During the course of the war, the Union lay siege to many of these fortifications, using all of their antebellum knowledge of ordnance and siege craft in ferocious artillery duels to effect their capture, such as that at Fort Pulaski. West Point graduates in gray resisted these attempts with equal reliance on military science.

Bases of operation on the land frontiers were also heavily fortified. Washington contained only one defensive work in 1860. Regular army engineers supported by retired officers—Professor Mahan included—helped to construct a true base of operations in the capital for the Army of the Potomac. The defenses of Washington eventually grew in circumference to thirty-seven miles, containing sixty-eight forts, ninety-three detached batteries, holding eight hundred guns and ninety-eight mortars, supported by twenty miles of infantry entrenchments and thirty-two miles of military roads. The base of operation at Cincinnati in 1862 included twelve miles of rifle trenches, four forts, and twenty-five detached batteries. Union defenses at St. Louis were only slightly smaller, while those of Nashville became the largest in the west. The base at Cairo had Vaubanian ramparts nine feet thick at their summit, twenty feet at the base, surrounded by a ditch twelve feet wide and ten feet deep. The Confederates built a honeycomb of similar fortifications along the Mississippi, on the Cumberland and Tennessee Rivers, along their coasts, and around Richmond and in northern Virginia. At Centreville the Confederates constructed impressive network of redoubts, lunettes, and angular bastions facing north and east to protect the base at Manassas

Junction. These, of course, were small compared with the elaborate network of field fortifications at Petersburg-Richmond in 1864–65, where soldiers used Mahan's prescriptions as a basis for improvement and innovation of fortifications design. As with Washington, Richmond constituted a political and cultural center of gravity and industrial hub. It was the enduring base of operations for the Army of Northern Virginia, explaining the scope and sophistication of the defensive works that surrounded the city.

Depots and bases of operation were only as good as their supporting transportation systems. Without links to other strategic points, they could only be relevant so long as supplies within them lasted. Therefore, leaders gave considerable attention to the establishment and maintenance of transportation networks. As in most things, the North was advantaged by existing railways, waterways, and roads. But the centralization of war making efforts also gave the Union an organizational advantage. Their department commanders could exert some influence over the use of roads and waterways. Meigs acquired and controlled tens of thousands of wagons and hundreds of boats, and maintained material and labor to build and repair roads, bridges, tunnels, and canals. Another West Point graduate, Col. (of Volunteers) Herman Haupt (class of 1835), ran the U.S. Military Railroad service, and it can be argued that few other individuals contributed more to Union success. The Military Railroad eventually controlled 2,300 miles of track, 400 engines, 6,600 rail cars, and employed dozens of thousands of men.[5] Ability to build and rebuild railways and rail bridges using prefabricated rail stock, engines, cars, and ingenuity made campaigning considerably easier for the Union. Lincoln is reported to have complimented him for constructing "the most remarkable structure that human eyes ever rested upon. That man Haupt has built a bridge across Potomac Creek [in seventy-two hours], about 400 hundred feet long and nearly a hundred feet high, over which loaded trains are running every hour, and, upon my word . . . there is nothing in it but beanpoles and cornstalks."[6] Resupply by rail and water gave great flexibility to armies maneuvering in vast territories. The Confederates never achieved the efficiency of Northern rail service, could not develop the means to manufacture railway equipment, and therefore lacked redundancy in railway use. This caused significant logistics problems for Confederate armies when abundance of food, forage, and equipment in one part of the South could not be transported to aid armies in need in other locations.

Railways became the primary means linking depots to bases and other strategic points upon frontiers. They also became—together with waterways—key lines of operation for advances forward of frontiers, and lines of communication between advancing armies and their bases. Discussion of possible strategic combinations by military and political leaders and civilian commentators used prewar terminology. Lines of operation and communication were favorites. It was natural for anyone thinking of strategic movements to consider the lines along existing roads, railroads, rivers, and canals. In July 1860 Maj. Gen. Irvin McDowell used the Orange & Alexandria Railroad as his line of operation to advance upon Centreville and attack P. G. T. Beauregard's flank at Manassas. Maj. Gen. Joseph Johnston reacted to this threat by moving forces to reinforce Beauregard along the line of communications provided by the Manassas Gap Railroad. McClellan and his Confederate opponents were restricted to the use of the Staunton-Parkersburg Pike as their lines of operation in the West Virginia campaign that same summer; a campaign wholly dedicated to Union retention of the vital Baltimore & Ohio Railroad, the critical line of communications between strategic points in the North. Maj. Gen. Carlos Buell concentrated his forces at a base at Louisville and used the Louisville & Nashville Railroad as a line of operation to seize a forward base at Nashville, believing, as Winfield Scott wrote, that "with Nashville for a [forward] base of operation the so-called Southern Confederacy could be effectively divided."[7] Grant moved from his base at Cairo using the Ohio and Tennessee Rivers as lines of operation to take Fort Henry and Donelson, and once occupied, these lines of operation became his lines of communication to the base at Cairo. Albert Sydney Johnston had various Confederate forces use the Mobile & Ohio and Memphis & Charleston Railroads to concentrate his army at Corinth to move upon Grant at Pittsburgh Landing. Indeed, it is impossible to understand the context of these campaigns and battles without knowledge of how these generals considered towns and hubs as strategic points and bases, and rail and waterways as key strategic lines of operation and communication.

During the campaigns of 1861–62, leaders and critics on both sides proposed strategies for winning the war. Scott had originally suggested a singular line of operations down the Mississippi, splitting the South, while at the same time blockading the entire coastline to coerce the Confederacy into rejoining the Union. Unfortunately for Scott, public criticism of this strategy only ended after his death when people began to realize the possible wisdom of his suggestion. McClellan, Halleck, Buell, and others offered advice to a grateful president on what strategic combinations might work best for offensive campaigns against the South. McClellan's were the most grandiose (and realistic), calling for sweeping land campaigns across the Confederacy, using incredible numbers of troops.[8] Except for Henry Halleck, these officers did not argue a preference for Jomini's interior lines. A slave to the theories that he wrote so eloquently of, Halleck felt compelled to lecture the president against all consideration of employing separate forces on exterior lines against concentrated Confederate forces. To be fair, Halleck's advice appears to have applied to strategy in a single theater and not across the continent, but even so, his reasoning was narrow. Lincoln, on the other hand, was consistently thinking on broader scales and stated in January 1862 that Union success would require the advance of separate field armies simultaneously against various Confederate frontiers in order to deny the enemy their "*greater* facility of concentrating forces upon points of collision."[9] Lincoln's reasoning reflected what most Civil War leaders appreciated: that no specific combination of interior or exterior lines granted a panacea. As time passed, it became more obvious that to defeat the South the Union needed to advance on several fronts at once and achieve a "concentration in time" by having disparate armies march deep into the Confederacy simultaneously.[10]

Debates about strategy in Richmond also revealed dilemmas. Most military thinkers favored a strategy of aggressive defense. This was an idea originally proposed by Vauban that had remained an essential component of Napoleonic warfare and the teachings of West Point. While strategically defensive, it called for limited attacks upon an enemy to spoil his offensive designs and "menace his lines of operation." West Point doctrine had presented this strategic option as "active defense."[11] Its adoption by the Confederacy, however, did not occur without controversy. Each state wanted to protect its territory against Union invasion with its own state troops, and with as much Confederate government support as possible. However, the limited resources at the disposal of the Richmond government were far too stretched to defend upon all frontiers equally, and to do so would leave no residual forces to launch the offensive operations called for in "active defense." The expanse of the Confederacy demanded a strategy placing field armies on critical frontiers and relying on the economy granted by fortifications and militias on other fronts.

One such other was New Orleans. Reliance on Third System defenses was justified at Wilmington, Charleston, Mobile, Pensacola, Galveston, and Savannah (even after Fort Pulaski fell). Richmond therefore felt that Third System forts at New Orleans would defend equally well and reduced manpower there in favor of concentrating in Tennessee. New Orleans fell once Adm. David Farragut ran past the Third System fortifications at night and seized a defenseless town. The concentration of forces in Virginia and Tennessee left such strategic points open to capture. At the same time, however, concentrated field armies gave the South a capability for offensive operations needed to keep the initiative, to capture Union supplies, to buy time in hopes of eroding Northern will to fight, and gaining British and French recognition and support. In this, they came close to success on more than one occasion.

The offensive mindedness of Northern strategists in 1862 did not translate into bold action by Union armies. McClellan, Buell, and Halleck proved to be deliberate if not glacially cautious, in stark contrast to the audacity displayed by A. S. Johnston, Jackson, and Lee—all of whom embraced antebellum prescriptions regarding celerity. Many Union commanders also failed to concentrate and engage most of their armies in battle, leaving the antebellum prescription about concentration of mass on a decisive point unrealized. McClellan, Pope, Burnside, and Hooker all failed to capitalize on their superiority of numbers and do what Lee almost always did—throw it all in during a fight. Yet, slow or audacious in maneuver, thrifty or extravagant with battlefield lives, the commanders of this war did plan the great maneuver campaigns of Tennessee, the Shenandoah, the Peninsula, Second Manassas, and Antietam using military science. Union and Confederate alike couched plans in terms of strategic points, bases of operations or supply, and lines of operation and communication. These campaigns really only gain full meaning when one appreciates this contemporary paradigm.

Campaigns of maneuver through terrain that was relatively undeveloped compared with Western Europe forced field armies to rely on expertise in topographical and field engineering. The use of Topog officers throughout the war was little different than their use in the antebellum period. Brig. Gouverneur Warren, chief topographical engineer of the Army of the Potomac, is perhaps the best known of the thirteen West Pointers who applied reconnaissance, field sketching, and map making skill in direct support of Union field armies. Union and Confederate commanders sent these Topogs on reconnaissance tasks ahead of advancing armies and employed them to lay out proper defensive lines. McClellan, Buell, Grant, Rosecrans, Lee, Johnston, Beauregard, Burnside, and others relied on engineer reconnaissance advice as Scott had in Mexico, and historians often overlook this aspect of military science in Civil War histories.

Until recently, they also have misread the role of field fortification and field engineering.

Historian Earl J. Hess has uncovered for us the extent and role of field fortifications used by armies throughout the war. Of 303 Civil War battle sites studied, he has found fortifications to be part of 213.[12] Hess has also disproven the widespread belief that the employment of field fortifications started with the Overland Campaign of 1864. In fact, both sides utilized field fortifications from the commencement of hostilities. The defenses at Centreville have been mentioned, but Confederate troops in West Virginia also used trenches and improvised defensive ramparts to slow McClellan's advance, and most 1861–62 Confederate strategic points in the west contained considerable field works. Lee's first order to the newly named Army of Northern Virginia in June 1862 was to dig in along a twelve-mile front before McClellan's army.

However, during this phase of the war, field armies seldom entrenched at night as a matter of routine (neither army did at Shilo, for instance). During the Atlanta and Overland Campaigns, nightly entrenchments were common, ensuring that most attacks in 1864–65 were upon some sort of field works. One Union soldier stated that in the summer of 1864 much of northern Georgia was "cut up by earthworks almost as thick as furrows in a ploughed field." So common was the practice that Sherman admonished Gen. George Thomas not to allow his soldiers to dig in every time they encountered the enemy for fear of losing the tactical initiative and momentum.[13] Hess believes that it was the psychological effect of continuous combat in close proximity to each other (and not the technical effect of rifled musket fire) that drove men to entrench whenever possible. Whatever the reason, when soldiers did throw up defensive fortifications, it was not West Point–educated engineers who planned and executed the bulk of this work but regimental and brigade officers who were following the design published in Mahan's treatise on field fortifications, reprinted multiple times in thousands of copies and in other printed works.[14] While volunteers were learning this tactical art, West Point commanders gained reputations for their strategic appreciation of fortifications. The *Richmond Examiner* stating in 1861 that "when West Point meets West Point, spade meets spade."[15] Field fortifications evolved from simple designs in 1861 to sophisticated trench systems in 1864–65, to a point where Grant opined that the proper entrenchments could practically defend themselves. The use of such works throughout the war, albeit to very different degrees, and their ready acceptance by volunteers, is evidence of confidence in prewar theory of field fortifications.

As with field fortifications, field engineering also affected how armies moved in open campaigns. Much of the eastern and western theaters were in wilderness or rough-hewn farmland. Roads were mostly rutted sand tracks, bridges were simple wooden structures, and many canals and "improved" waterways were often small shallow ditches. The movement of dozens of thousands of men and hundreds of tons of material and supplies along roads and waterways required engineering skill in road, bridge, and canal construction.[16] Both sides began early in the war to employ engineer officers and large civil labor forces to improve avenues of transportation. Approximately twenty-three West Pointers served as brigade and division chief engineers, and thirty-four were corps and army chief engineers (with another fifty-five serving as district and departmental engineers). Buell used such men to build bridges in his advance upon, and from, Nashville. Pope dug a shallow canal to bypass Island No. 10 in his Mississippi operation in April 1862. McClellan used special pontoon trains to allow him flexibility to cross the Potomac where and when he willed, and he improved roads during his advance up the Peninsula. Joe Johnston at the time complained that "McClellan will adhere to the system adopted by him last summer and depend for success upon artillery and engineering. We can compete with him in neither."[17] Haupt employed one of his "wrecking and construction" trains containing prefabricated bents and stringers and construction gangs to reopen the Orange & Alexandria Railroad during the Second Manassas Campaign. The use and employment of civil engineering skill in field engineering tasks only increased in scope and sophistication as the war progressed. Much of the success of Grant's 1863 Vicksburg campaign and Sherman's Atlanta and Carolinas campaigns depended on engineering expertise. By 1865 field armies had become incredibly agile in maneuver because of the work of Topogs and officers who knew how to throw up bridges and improve roads or dig a canal. Grant demonstrated this in front of Lee when his army constructed a bridge across the James River (700 yards wide) in only seven hours on 14 June 1864. A West Point education proved as valuable in the Civil War as it had in Mexico.

Good logistics often determined campaign outcomes, and logistics planning was an integral part of strategic planning. The traditional methods of supply used in the antebellum period were still employed during the Civil War: supplies flowed from depots and industries through central logistics hubs out to field armies; regimental, brigade, and division quartermasters also bought supplies through local purchase, just as soldiers bought things from camp sutlers; and armies confiscated supplies in occupied territory by foraging. All three systems continued throughout the war. However, the only system that could guarantee supplies for large field armies was centralized logistics. This forced armies to create advanced bases of operation through which supplies could flow forward or upon which armies could retreat to reconstitute if required. In 1862 and 1863, the Union developed forward bases at Nashville, Memphis, and New Orleans and at Fortress Monroe and Aquia Creek Landing on the west bank of the Potomac northeast of Fredericksburg, Virginia.

In this last location, Union forces improved the landing area with docks and storehouses and the roadway and railhead leading south and created one of the busiest ports in the United States in order to resupply the Army of the Potomac during the winter of 1863. Every day 220 tons of commissary stores, 90 tons of quartermaster stores, and 440 tons of forage flowed through this base of operations to the corps locations on the Rappahannock line.[18] Such quantities required the ability to stockpile and a means to transport stock, particularly when armies used bases for offensive operations. In addition to the classes mentioned above, 20 tons of ammunition and 20 tons of railroad supplies also came through the Aquia base each day for the offensive operations in 1863. Equal scales of supply flowed into Chattanooga in late 1863 and 1864 (by rail and road from Nashville and from supplies coming upriver on the Tennessee). Road, railroad, and waterway lines of communications constituted the only options for the effective supply of field armies. Sherman paid particular attention to them on his advance to Atlanta. The *Official Record of the Rebellion* and autobiographies of the principal commanders reveal hundreds of references to "base of supply," "base of operation," or simply "base" as well as lines of operation and commmunciation.[19] Bases were almost invariably located at transportation hubs, strategic points that connected lines of communication with forward lines of operation. At the end terminus of a logistic rail or water line, stores were transferred from rail cars or boats to wagons and horseback. Regardless of the sophistication of these lines, the final distribution of stores, food, and ammunition was by the age-old method involving horse and muscle power. Generals desired to make this portion of the logistics chain as short as possible.

In 1862, after the capture of New Orleans, the Union used the city as a base and began to extend a new line of operations up the Mississippi River, capturing Baton Rouge. Simultaneously, Buell attempted to march from Corinth to Chattanooga along the line of the Memphis & Charleston Railroad, but his single line of communication from the Louisville depot through bases of operations at Nashville and Corinth proved much too long and vulnerable. Grant later attempted to advance from his base at Columbus to take Vicksburg. He initially chose the Mississippi Central Railroad as his line of operations, but this single line extended too far from

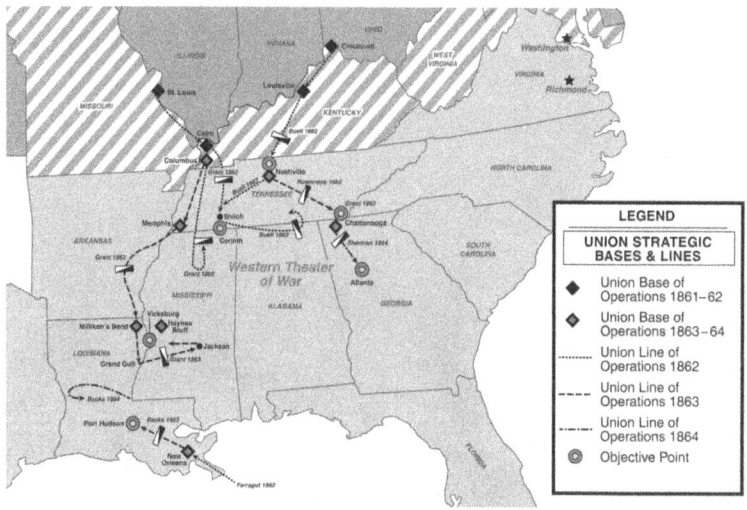

MAP 9. Western theater of operations.

his base and became susceptible to interception by raiding Confederates under Generals Earl Van Dorn and Nathan Bedford Forrest. In 1863 Grant chose a line down the Mississippi River using Columbus and Memphis as bases, trying various expeditions and engineering feats (including construction of a bypass canal) to put himself in position to attack the city. Eventually he was able to use the temporary forward base at Milliken's Bend to build an expedient road and project forces south of Vicksburg on the west side of the Mississippi, then cross and strike east of the city, cutting off the Vicksburg garrison from its line of communication. In this maneuver, Grant severed his line of communication from the base at Milliken's Bend, moved independently to capture Jackson, and then reestablished communications through a new base at Haynes Bluff (northeast of Vicksburg). He then invested the city and forced its surrender after a siege. Grant's creative intelligence was unconstrained by pedantic notions regarding interior versus exterior lines of operation, and he was not afraid to switch from one base to another during a campaign.

FIG. 7. (*Above*) Confederate fortifications near Centreville, Virginia, March 1862. Courtesy of the Library of Congress.

FIG. 8. (*Opposite top*) Petersburg, Virginia. Confederate fortifications with *chevaux-de-frise* beyond. Courtesy of the Library of Congress.

FIG. 9. (*Opposite bottom*) City Point wharf with Federal artillery train, 1864–1865. Courtesy of the Library of Congress.

FIG. 10. Yorktown, Virginia. Embarkation point for White House Landing. Courtesy of the Library of Congress.

FIG. 11. Pontoon bridge on the James River, 1864. Courtesy of the U.S. Department of Defense, Brady Collection.

To be fair, McClellan also proved to be creative with the concept of the base. At the theater strategic level, he devised the use of alternate bases of Washington and on the Peninsula to provide him options to maneuver against the Confederates in Virginia. During his Peninsula campaign, he displayed mental agility in first using Fortress Monroe as his base, then switching to a base at White House on the Pamunkey (York) River. This base would have allowed him to use the Richmond and York River Railroad for deployment of large siege guns when ready to besiege Richmond. However, Lee foiled his plans by executing a flank attack on McClellan's right, severing the Army of Potomac from its White House base and forcing McClellan to open a new base at Harrison's Landing on the James. Unfortunately, the Harrison base contained no railhead and could not meet McClellan's campaign needs.

Lee's Gettysburg campaign also tested the theory of the base. Driven by a perennial need to acquire forage, food, and military supplies directly from Union stocks and the relatively rich countryside north of the Potomac, Lee advanced from a base in the Shenandoah in 1863 into Pennsylvania bent upon a long summer's campaign. Intent on foraging supplies, he nevertheless retained a line of communications to Shenandoah. While the ensuing battle of Gettysburg appears to be well trodden by historians, relatively little analysis exists regarding Lee's planning and conduct of campaign logistics.

On the other hand, the Union advance from Nashville to the capture of Chattanooga that same summer is a well-studied demonstration of the growing maturity of Union war making efforts. Few generals of the war were as capable as was William Rosecrans in employing military science during an offensive campaign. His use of multiple lines of operation from the Nashville base, balanced corps structures, reconnaissance, and field engineers capable of remarkable feats of bridge building dislocated Confederate efforts to defend in Tennessee. However, Rosecrans, like McClellan, proved better at planning and maneuver than fighting, and he lost what momen-

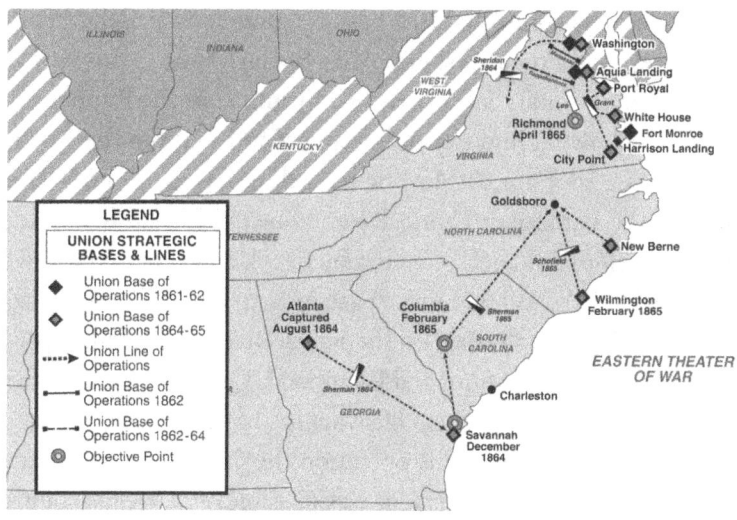

MAP 10. Eastern theater of operations.

tum he had at Chickamauga. Sherman proved to be better rounded, combining both superb planning with fighting skill in his 1864 overland march on Atlanta. He used a broad-front advance to protect his single line of communication on the Western & Atlantic Railroad, keeping his armies linked to the Union forward bases at Chattanooga and Nashville, which were in turn linked to depots at Louisville and Cincinnati. Careful planning and skilled maneuver kept Sherman's armies strong during this extended advance. His reconnaissance and integral field engineering capabilities were equally robust, retaining the ability to rebuild bridges and repair rail lines with remarkable speed. One Confederate despaired at the impotence of a plan to destroy a tunnel in front of Sherman's advancing columns, stating that Sherman likely carried a spare tunnel in his trains.

Grant's 1864 Overland march demonstrated different but equally mature expertise in campaigning. Grant did what Sherman could not, switch bases of operation and swivel around the enemy's flank with great dexterity. From the commencement of the advance in early May 1864 until after the battle of Spotsylvania, the Army of the Potomac was connected to a base of operations at Belle Plain, Aquia Creek, and Fredericksburg. After Spotsylvania, Grant moved his base to a new point at Port Royal, Virginia, on the Rappahannock River, allowing him to maneuver east and south to the North Anna. A week later (on 28 May) he moved his base again to White House on the Pamunkey (York) River, allowing him to maneuver east and south to fight at Totopotomoy Creek and eventually Cold Harbor. He left the White House base in mid-June and connected to a different base established on the James River, facilitating a partial investment of Petersburg and Richmond. Grant understood the value of the base and the benefit of switching bases to enable maneuver. Historians later called his operation the Overland Campaign, but in fact his overland movement was possible only because of his use of maritime lines of communication and shore-side bases of operation (just as he had done in Mississippi).

Lee complained of Grant's Overland Campaign that his "gradual whirl and change of base from Fredericksburg to Port Royal, then to York River and then to James River, as a thing which, though foreseen, it was impossible to prevent."[20] Grant had wanted to destroy Lee in a grand "decisive" battle during this campaign, but in failing to do so, he used maneuver facilitated by multiple bases of operation to fix Lee. Once Grant forced Lee into trenches around Richmond, he sought to sever the connection of Richmond to depots and other Confederate bases west and south. He asked Sheridan to destroy the Shenandoah as an alternate base, then to cut the Virginia Central. Gen. George Stoneman was to cut the Virginia & Tennessee Railroad, and eventually Grant directed Generals Warren and Sheridan to cut the South Side Line, eliminating the chances of Lee moving to an alternate base of operations in the valley or in North Carolina (linking up with Joe Johnston's gathering army).

In the western theater, things were as interesting. After Sherman took Atlanta in 1864, he had to contend with Gen. John Bell Hood's army attempting to sever the Union line of communication to Chattanooga and Nashville. In the open spaces of North Geor-

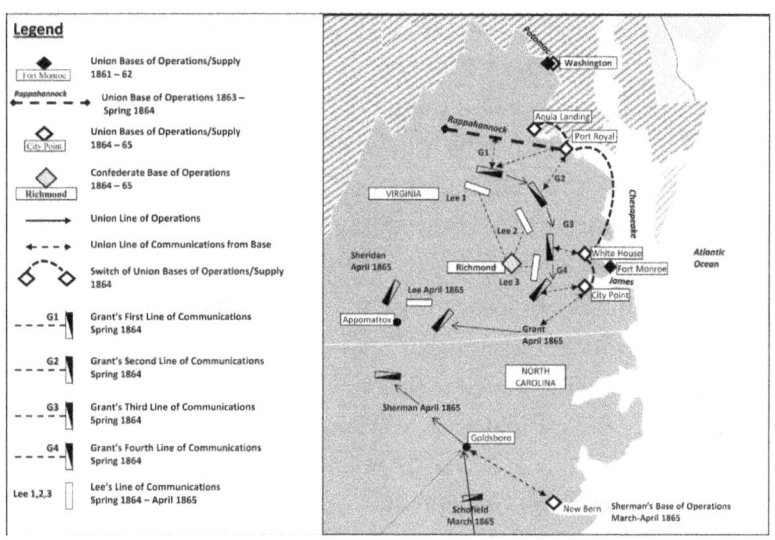

MAP 11. Grant's use of bases of operation, 1865.

gia, Sherman could not pin Hood down to fight, nor could he leave his line open to such attacks. By this time he realized the impractical nature of Atlanta as a forward base of operations, so distant from the depots that sustained it. He decided then to ignore the Confederate Army of the Tennessee as it went into Alabama and instead cut his own line of communications from the Nashville/Chattanooga base and marched to Savannah on the coast, disrupting and neutralizing the heartland of the Confederacy en route. At Savannah Sherman could connect to another base of operations. During this march, he lived off what he carried from Atlanta and what he confiscated, and he destroyed anything of military or economic value to the South. When he reached and captured Savannah, he found that Quartermaster General Meigs had stockpiled sufficient uniforms and boots and all other classes of supply to totally reequip Sherman's 60,000 soldiers.

Sherman then began a campaign to dismantle the Confederate's eastern depots and bases, marching his army through the Carolinas, destroying military and economic material. His march contended with little Confederate resistance but with considerable issues of weather and wet, swampy terrain. Sherman was undeterred, building roads and bridges in advance of his moving columns to the complete surprise of his enemies. Unable to sustain his advance from a base of operations in Savannah, he used maritime freedom of movement to switch to another base established after the capture of Wilmington and New Berne in February 1865, receiving yet another issue of clothing and stores from Meig's well-oiled logistics machine. He then prepared either to confront Joe Johnston's newly formed force near Goldsboro or to march to join Grant. Meanwhile, Grant's pressure at Petersburg-Richmond, combined with the ever-tightening noose around Confederate rail lines supporting the Army of Northern Virginia, left Lee one option—an attempt to move southwest into the Carolinas. However, Sherman was at this time making the southeastern interior of the Confederacy unsuitable as an alternate base of operations for the Army of Northern Virginia, and Lee's attempt ended with his army trapped at Appomattox in April 1865.

In conducting the 1864 and 1865 campaigns, the Union was applying the science of military movement. The governing paradigm for planning and conducting campaigns was the use of the base (tied to depots) and lines of operation to seize strategic objective points that provided advantage over Confederate forces.

Antebellum emphasis on logistics planning combined with industrialization in society allowed for Northern production and stockpiling of military materiel on grand scales. Union leaders struck a balance in manpower usage, acquiring adequate recruits for battle while keeping sufficient civilian workers to sustain the war economy, while in the South leaders took from society enough men to preclude effective war production but not enough to win decisively in battle. Inadequate war planning and appreciation of logistics took the South to a point where insufficient labor caused poor food and equipment production and distribution, which in turn contributed to considerable absenteeism in the armies. The deteriorating situation in 1865 forced the Confederacy to consider not only conscription of the slaves but also their emancipation. Emphasis on innate Southern manliness, courage, and martial genius overshadowed acknowledgment of the importance of military science, something that, together with the application of states' rights doctrine to war making, helped bring ruin to the Confederacy.

Unafraid of advances in technology and inculcated in the spirit if not the practice of engineering, Union commanders and staffs organized civil, topographical, and field engineer and ordnance support to their field armies to an extent unachieved in the South. These officers linked the mobilization and industrial base of the Union to the battlefield as part of a continuum. And in strategy, too, the North evolved. Grant and Sherman's appreciation of strategic movement developed between 1862 and 1865 to understand that the application of the theory of bases of operations was fundamental in war, and multiple options for bases of operations facilitated maneuver against an enemy in *la grande guerre*. Had West Pointers been as equally comfortable with *la petite guerre* as an alternative to conventional warfare, the war might have dragged on after Appomattox. However, the dominant paradigm of war of this era, represented in antebellum military science for *la grande guerre*, discouraged this.

This is not to suggest that the application of military science guaranteed Union victory. Until Lincoln was reelected in November 1864, Union political resolve to continue the war was in serious doubt. These contingent factors outweighed sheer Union military might, and until the end, Lincoln was never certain about the outcome. Only in the spring of 1865, when Union armies denied Confederate forces effective bases of operation from which to muster, project, and supply military forces, did military defeat become inevitable.

If there was a real advantage to the Union during these years of war, it was in the application of military science by well-educated and trained minds. While historians rightfully credit the genius of Lincoln, Grant, and Sherman with the ultimate military success of the North, the conditions of success were laid by a large number of professional officers who shared a paradigm of how to prepare military frontiers, fortify strategic points, identify ordnance requirements, muster, equip, arm, train, and move large volunteer forces, properly encamp and supply these forces, and plan for their employment in successful offensive campaigns. As much as military genius, the Union's advantage in a shared paradigm of military science proved influential. The legacy of the Civil War in the U.S. Army was of continued faith in staff planning, logistics, engineering, artillery competency, and integration of the means for industrial warfare and a reliance on volunteers. Theirs had become a scientific way of war.

Conclusion

Inheritors of an unscientific military tradition that emphasized the role of the amateur in irregular warfare, Americans entered the Revolution in want of expertise in the scientific disciplines of artillery and fortifications. Contempt for standing armies had also prevented colonial leaders from appreciating the problems related to raising and supplying military forces for campaigns. Events of the Revolution revealed the ignorance. Yet the desire of each state to remain free of financial obligation to the federal government guaranteed lukewarm support for the maintenance of the Continental Army and no sooner had hostilities ended in 1783 than Congress began efforts to expunge what had been learned during the war. In the ensuing years, the militia myth bolstered opposition to a large military establishment dedicated to national preparedness for war. States and non-Federalists remained suspicious of military forces that might suppress political opposition or coerce states to bend to the federal will. However, the issue was as much economic as political or philosophical. The states remained reluctant to shoulder the costs of national defense. This sentiment changed somewhat in Atlantic states with the threat of war in 1794, 1798, and 1806, and real war in 1812, when they agreed to federal plans for the construction of permanent coastal fortifications. Such plans required expertise in military science, providing impetus for the establishment of the U.S. Military Academy at West Point in 1802, with its

CONCLUSION

mandate to educate "scientific officers" capable of building the fortified maritime frontiers.

During the Jeffersonian period, states agreed to pay considerable sums of money for defensive infrastructure, in lieu of a large standing army. President James Monroe and Secretary John Calhoun attempted after the War of 1812 to convince Congress of the need for more troops, but could secure support for only 6,000 regulars. Throughout most of the antebellum era, the size of the army remained between 6,000 and 13,000 soldiers, fluctuating during periods of conflict in Florida, in Mexico, and along the western and northern frontiers. The state volunteer system, supported by an enduring narrative of the natural effectiveness of the American militiaman, obfuscated the need for a larger army even though few states spent the funds needed to maintain a well-regulated militia.[1]

Accepting the lack of political appetite to increase the number of regulars, Monroe and Calhoun turned instead to ways to increase military capability by professionalizing an "expansible" army. By 1820 they had succeeded in crafting enduring defense policy, establishing staff bureaus, publishing regulations, and institutionalizing a professional curriculum at West Point. These significant achievements had lasting impacts. Calhoun's defense policy appealed to states' representatives as it committed resources to the development of fortifications along the coastal and land frontiers and corollary improvements of commercial lines of communications internal to the frontiers. This expedient system of national defense addressed opposition to large armies while at the same time guaranteeing a basic level of military funding and support for the remainder of the antebellum period. It also created demand for military science. West Point's largely French curriculum met the demand by delivering a "complete system" that supported the nation's defense policy. Its core was a doctrine for *la grande guerre.*

West Point cadets received a sound military education that distinguished them from officers who had received commissions directly from the secretary of war and from the state-appointed officers of militia or state volunteers. At least until state-run military schools started in the 1840s, West Point offered what no other school could—a core curriculum of mathematics, physical sciences, and engineering. No other school offered the blend of training in all branches of military service together with this scientific curriculum. The academy produced an annual class of fifteen to thirty graduates who steadily came to fill most of the commissioned officer positions of the regular army.

However, while it was unique, West Point was not out of place in Jeffersonian America. Desire for a national institution of science was a dream of Washington and Jefferson. The demand in society for skilled surveyors, civil engineers, architects, and mapmakers generated widespread political support for the academy's curriculum. It is true that Americans periodically criticized this for its narrow scope, with annual boards of visitors suggesting curriculum expansion to include instruction in classical languages, literary arts and criticism, rhetoric, dialectics, law, and history. However, throughout the entire antebellum era the academy's leaders insisted that these subjects were "accessory" and potentially disruptive to the focused study of science, a reflection of a deliberate preference for Jeffersonian Enlightenment thinking. But this sentiment also represented an American reality.

Throughout the antebellum period, neither specialization of military function nor a generalized liberal arts education served the needs of a small U.S. Army, spread thinly in small outposts over a vast continent. More than anything else, each graduate of the military academy needed some skill in performing local staff functions, in organizing local depots, in surveying, in building defensive fortifications and garrison infrastructure, and in constructing roads, canals, and bridges. Such skill could only come from a scientific curriculum. Each graduate was also eligible for detached service with one of the staff departments where knowledge of ordnance, supply, and quartermaster functions was essential.

CONCLUSION

The last half of this work has explained that satisfying civilian demand for engineers was not the purpose of West Point and was not part of the planning in the War Department. While the U.S. Army did attempt to assist in federal and state civil engineering, military preparedness and the conduct of operations occupied most of the effort of the army. The military science learned at West Point helped the nation's internal improvement, but it was also fundamental to officers performing a wide array of purely military duties. West Point graduates maintained a versatile set of skills; indeed, they were military generalists, equally capable of company and battalion leadership, of organizing, training, moving, and supplying battalions of volunteer state troops and performing duties as general staff officers directly for the War Department.

In the military reforms enacted after the War of 1812, the creation of a military general staff in Washington stands out as one of the most important. This included bureaus under a quartermaster general, a commissary general, an adjutant general, a surgeon general, a paymaster general, a chief of ordnance, a chief of engineers, and a chief of topographical engineers. The bureaus and their chiefs reported directly to the secretary of war and were responsive to, and reliant upon, Congress. The factor that distinguished this general staff from continental European models (especially that of the later widely acclaimed Prussian general staff) was its nearly complete control by political masters and not by commanding generals. In a young republic suspicious of concentration of power, the separation of staff from command functions and political oversight remained a chief characteristic of the emerging American military system. Sometimes much maligned, this system worked well for the United States until late in the twentieth century.

The military science of West Point was essential for the smooth running of the new American staff system. The qualifications of the graduates pretty much guaranteed their domination of the staff bureaus in Washington and in satellite offices in all geographic military departments and districts and posts. At any time during the antebellum period, the War Department detached a significant portion of the officer corps (between 22 and 35 percent) on service of the general staff bureaus. In this manner, well over 40 percent of West Point graduates (and a majority of long-service graduates) gained considerable staff experience during times of peace. This in turn helped to professionalize the army and provide valuable experience to young officers that became useful to them in times of war.

During the Seminole War of 1835–42, the various expeditions against Indian tribes, and during the Mexican War, the staff and line experience of the regular officer corps contributed to military success. Particularly noteworthy was the efficacy with which West Point graduates worked as quartermasters, assistant adjutants general, commissariat officers, ordnance officers, paymasters, reconnaissance officers, and engineers to facilitate the raising, organization, training, equipage, movement, and supply of volunteer and militia forces. These officers, mostly from infantry and artillery units, could serve over time in several staff capacities and sometimes in command of troops from other branches. Their varied employment throughout the period 1820 to 1860 demonstrated versatility. That they regularly dealt with issues of supply and administration across vast distances, using a mix of military and civilian contracted resources, without problems of accountability to Congress or military failure, is significant. They constituted a general staff capable of overseeing the expansion of American military forces when needed.

For these reasons, West Point military science proved politically and socially acceptable. The academy itself came under periodic criticism from various factions in Congress, from state politicians who would have preferred to see money spent on state militias, and from antiprofessionals who occasionally accused the academy of elitism. These attacks ceased to have meaning after the victory by the U.S. Army in the Mexican War. By the eve of the Civil War, military science had become the dominant paradigm for those Washington-based officers who shaped defense policy, their colleagues who taught at the U.S. Military Academy, and the graduates.

CONCLUSION

It is no accident of history that cannons from Third System fortifications at Charleston Harbor fired the opening shots of the Civil War, in a duel between a former West Point artillery instructor (Major Anderson) and one of his students (Brig. P. G. T. Beauregard). Neither is it coincidental that the last great battle of the war was a grand siege at Petersburg and Richmond, involving a greater extent of entrenchments and fortifications than had previously ever existed. The war ended in a real sense when a Jeffersonian West Point graduate (Robert E. Lee) surrendered to a late–Jacksonian era graduate (Ulysses S. Grant). Lee was a man of science, who served without sentimental trappings. Grant was similarly devoid of the romanticism, especially of the variety that marked the Jacksonian era. He was more representative of the Realism that was to follow the Civil War. If there was a connection between these two, it was their West Point military science, and their differences explain two distinct phases of that science (antebellum and postbellum). In between Sumter and Appomattox were four years of war that drew on all of the elements of the antebellum academy curriculum. West Pointers used fortifications, logistics, engineering, advanced artillery applications, and strategy, largely as taught and understood in the antebellum period. New techniques and technologies changed the scope of application of most elements. The railway and telegraph had a profound impact on logistics and command and control. Engineering capacities expanded in scope to meet the demands of railway, road, and water troop movements and to conduct the many sieges of the war. Artillery technology and techniques improved as officers learned to mass artillery on the battlefield and to defend and attack fortified positions with more advanced applications of fire. Field fortifications changed the nature of Civil War tactical battles from something resembling Napoleonic warfare to something altogether different, very much resembling engagements that would reappear in the First World War. In all this change, the theoretical basis of military science, and strategic movement in particular, changed not at all. Advances in technology reinforced the importance of proper selection of strategic points and lines.

CONCLUSION

The Civil War propagated knowledge about logistics, ordnance, fortifications, and administrative preparedness and redundancy. The fundamental principle remained: war required the movement of massed forces from secure bases of operation and their concentration upon decisive points. The Civil War diffused military science so well that it became the way of war for the U.S. Army for the next century.

The academy's curriculum between 1820 and 1860 had reinforced the Jeffersonian notion of science as key to all human affairs. With the Civil War, this began to change significantly. Appreciation for evolution and realistic criticism of Jeffersonian science and Jacksonian romanticism alike, slowly replaced ideas of a perfectibility of war. Industrialization, urbanization, continental settlement, and the end of the western frontier produced change throughout American society, and West Point was no exception. Yet, however bothered academy professors were by these changes, the academy's military science was safe. Commitment to a science of war that taught immutable principles, based on appreciation of topography and military mass, portrayed in geometric form, remained in the U.S. Army for many decades to come, especially when Sherman served as General of the Army. His Superintendant of the West Point academy, Maj. Gen. John Schofield, used Sherman's Georgia campaign, with his switching of bases of operation from Atlanta to Savannah, as a model application of the science of war.[2]

Bvt. Col. J. B. Wheeler (class of 1855) replaced Mahan as professor of engineering and the science of war after Mahan's death in September 1871. In 1893, he published his own teaching notes in *A Course of Instruction in the Elements of the Art and Science of War* within which he reiterated the ideas of military science laid down by Mahan before him. Wheeler stated that the science of war was to be studied in the same way as other sciences, with attention paid to its guiding principles. The principles were the result of experience and analysis of that experience. One should study past military operations in order to "separate the principles from the details. . . . The theories of the science are therefore based upon the acted past

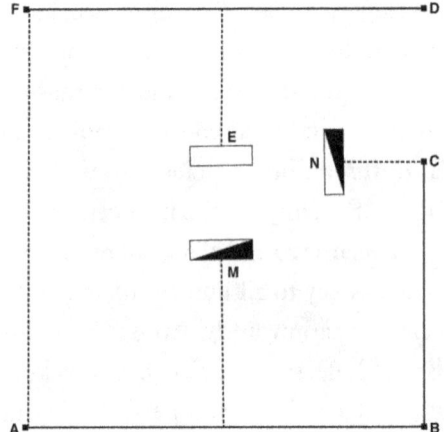

FIGURE 12. Wheeler's concept of strategic movements, ca. 1890. Based on *A Course of Instruction in the Elements of the Art and Science of War for the use of the Cadets of the United States Military Academy* by J. B. Wheeler (New York: D. Van Nostrand, 1893), 16.

and not upon the imagined future."[3] Wheeler read his Clausewitz and was aware of the contemporary American admiration for the German general staff. However, such influences were not changing traditional American military thinking. Acknowledging the role of "chance" in war, Wheeler suggested that a master of the art of war should study military science to be able to "eliminate the elements of 'chance' from any military operation."[4]

In planning and conducting campaigns using the immutable principles of the base of operations, lines of operation, and strategic objective points, Wheeler urged his cadets to study and employ the latest in railway, signals, and artillery applications. Wheeler's science of war did not forget fortifications. By 1890 this subject had been separated from strategy in the West Point curriculum, but it was no less important. Wheeler taught field and permanent fortifications using more recent texts, but these were very much Mahanian and therefore Vaubanian in content. At the turn of the century, the course in the science of war began to incorporate more historical analysis and an annual visit to one of the battlefields of the Civil War. The elements of instruction throughout this curriculum continued to emphasize the importance of impersonal forces in preparing for and conducting war. The principles of mobilization of state volunteers and the theory of the base, the importance of fortifications remained essential. Permanent coastal defenses, although transformed by technological advances, remained the singular consistent element of American defense strategy for another fifty years. The U.S. Army continued to teach and practice field fortifications, confirming their importance during the slaughterhouse of trench warfare on the western front in 1917–18.

Historians have debated whether the Civil War was the last Napoleonic war or the first modern war. Either view sees it as a dividing event. My research and my view of history as *la longue durée* tell me that the Civil War was really part of a continuum and that military science explains better the place of the Civil War in military history. This elevates the importance of understanding the role of doctrine within organizational culture before wars begin and its adaptability in war. Components of antebellum military science remained evident in the way of war of the United States into the twentieth century, in mobilization for the First World War, and the methodology used in planning and executing Second World War campaigns. Of the alternatives to war preparation referred to in my introduction—a reliance on amateur citizen soldiers in states' militias, a preference for *la petite guerre*, or a large elite-run standing army—the United States avoided all of these until the second half of the twentieth century, when it replaced this small nineteenth-century model with the last alternative. However, the intellectual approach to war, evidenced as late as Operation Desert Storm, showed continued preference for science over intuition and genius. It is only within the last decade, when proper understanding of the fundamental importance of the base of operations has been obfuscated by delusions of Clausewitzian centers of gravity, that the U.S. Army has forsaken the enduring nineteenth-century scientific principles that contributed so much to American success.

This book has examined a force of continuity in American military thinking for the period 1820 until the Civil War and suggests a continuity of thinking thereafter. Specifically, it has analyzed what constituted nineteenth-century American military science, why it was framed within government policy and taught within the U.S. Military Academy, and how it became the early American way of war. It refutes two dominant schools of revisionist thought: the first regarding how Swiss military theorist Antoine de Jomini dominated antebellum military thinking and the second regarding the lack of military professionalism in the antebellum army. It explains instead that the doctrine of military science prevalent in the U.S. Army in the antebellum period originated not from Jomini but from an enduring military enlightenment that had been transplanted from Europe during the early republic and Jeffersonian eras, and thereafter formed the basis for a commonality of thought among officers who were professional in their preparations for war in 1861. Theirs was a way of war that conserved Enlightenment ideas. They believed strongly in the possibility of acquiring a perfect knowledge of war. While incorporating distinct branches—fortifications and engineering, artillery applications and ordnance functions, logistics, and the science of strategic movements—all were related by a common reliance on mathematics. Mathematics distinguished this doctrine from previous ones and offered a basis for decision making beyond intuition. Faith in this Enlightenment idea was waning elsewhere by 1850. However, at West Point and in the minds of academy graduates serving throughout the union, Enlightenment thinking continued, asserting that war was governable by a grand theory, necessitating the comprehensive study of military science. In many respects, this thinking lives on in the U.S. military to this day.

APPENDIX OF TABLES

TABLE 1. Regulations of the U.S. Military Academy at West Point, New York, 1823

Time	4th Class	3rd Class	2nd Class	1st Class
From dawn of day to sunrise	*Reveille at dawn of day—Roll call immediately after Reveille—Police of rooms—Cleaning of arms, accoutrements, etc.—Inspection of rooms thirty minutes after roll call*			
From sunrise to 7 o'clock	Study of Mathematics	Study of Mathematics	Study of Natural and Experimental Philosophy	Study of Engineering and Military Art
From 7 to 8	*Breakfast at 7 o'clock, guard mounting at half past 7— Class parade at 8*			
From 8 to 11 o'clock	Recitations in Mathematics	Recitations in Mathematics	Recitations in Natural and Experimental Philosophy	Recitations and Drawing relative to Engineering and Military Art
From 11 to 12	Study of Mathematics	Study of Mathematics	Lectures on Natural and Experimental Philosophy	Lectures on Engineering and Military Art
From 12 to 1 o'clock	Study and Recitation of French	Study and Recitation of French	Study or Lectures on Chemistry Alternating	Chemistry or Mineralogy & Geology Alternating
From 1 to 2	*Dinner at 1 o'clock—Recreation from dinner to 2 o'clock*			
From 2 to 4 o'clock	Study and Recitation of French	Drawing Figures or Study of French Alternating	Drawing of Landscape and Topography	Study and Recitations of Geography, History, Ethicks, and National Law

4 to sunset	Military exercises—Dress parade and roll call at sunset			
From sunset to ½ hour past	Supper immediately after parade—Signal to retire to quarters immediately after supper			
From ½ past sunset to ½ past 9	Study of Mathematics	Study of Mathematics	Study of Natural and Experimental Philosophy	Study of Engineering and Military Art
From ½ past 9 to 10 o'clock	Tattoo at half past 9 o'clock—Roll call immediately after tattoo Signal to extinguish lights, and inspection of rooms at 10 o'clock			

APPENDIX OF TABLES

TABLE 2. Regulations of the U.S. Military Academy at West Point, New York, 1832

Times of Day	4th Class	3rd Class	2nd Class	1st Class
From dawn of day to sunrise	*Reveille at dawn of day—Roll call immediately after reveille—Police of rooms—Cleaning of arms, accoutrements, etc.—Inspection of rooms thirty minutes after roll call*			
From Sunrise to 7 o'clock	Study of Mathematics	Study of Mathematics	Study of Natural Philosophy	Study of Engineering and the Science of War
From 7 to 8	*Breakfast at 7 o'clock, guard mounting at half past 7—Class parade at 8*			
From 8 to 11 o'clock	Recitations in Mathematics	Recitations in Mathematics	Recitations in Natural Philosophy	Recitations and Drawing relative to Engineering and the Science of War
From 11 to 12	Study of French	Recitation and Study of French	Recitation and Study of Chemistry	Lectures on Engineering and Military Art
From 12 to 1 o'clock	Study of French	Recitation and Study of French	Lectures and Study of Chemistry	Study of Rhetoric and Moral and Political Science, and Mineralogy and Geology
From 1 to 2	*Dinner at 1 o'clock—Recreation from dinner to 2 o'clock*			
From 2 to 4 o'clock	Recitation and Study of French	On alternate days Drawing of the Human Figure	Drawing of Landscape and Topography	Recitation of Rhetoric and Moral and Political Science, and Mineralogy and Geology
From 4 to sunset	*Military exercises—Dress parade and roll call at sunset*			
From sunset to ½ hour past	*Supper immediately after parade—signal to retire to quarters immediately after supper*			
From ½ past sunset to ½ past 9	Study of Mathematics	Study of Mathematics	Study of Natural Philosophy	Study of Engineering and the Science of War
From ½ past 9 to 10 o'clock	*Tattoo at half past 9 o'clock—Roll call immediately after tattoo Signal to extinguish lights, and inspection of rooms at 10 o'clock*			

APPENDIX OF TABLES

TABLE 3. Regulations of the U.S. Military Academy at West Point. New York, 1856

Times of Day	5th Class	4th Class	3rd Class	2nd Class	1st Class
Dawn to sunrise	*Reveille at dawn of day—Roll call immediately after reveille— Police of rooms—Cleaning of arms, accoutrements, etc.—Inspection of rooms thirty minutes after roll call*				
Sunrise to 7 o'clock	Study of Mathematics	Study of Mathematics	Study in Natural and Experimental Philosophy	Study in Civil Engineering	Study of Engineering and the Science of War
From 7 to 8	*Breakfast at 7 o'clock, guard mounting at half past 7—Class parade at 8*				
From 8 to 11 o'clock	Recitations in Mathematics	Recitations in Mathematics	Recitations in Natural and Experimental Philosophy	Recitations in Civil Engineering or Moral Sciences	Recitations in Engineering and the Science of War
From 11 to 12	Study of Mathematics	English studies or Fencing Alternating	Study and Recitation in French and Spanish Alternating till January then Spanish	Study of Chemistry or Riding Alternating	Mineralogy & Geology, Ordnance & Gunnery, Riding, Practical Engineering Alternating
From 12 to 1 o'clock	English studies or Fencing Alternating	Recitation and Study of French		Study of Chemistry or Riding Alternating	
From 1 to 2	*Dinner at 1 o'clock—Recreation from dinner to 2 o'clock*				
From 2 to 4 o'clock	English studies or Fencing Alternating	French	Drawing and Riding Alternating	Drawing	Law, History & Conduct
4 to sunset	*Military exercises—Dress parade and roll call at sunset*				
Sunset to ½ past	*Supper immediately after parade—signal to retire to quarters immediately after supper*				
From ½ past sunset to ½ past 9	Study of Mathematics	Study of Mathematics	Study in Natural and Experimental Philosophy	Study of Civil Engineering or Moral Sciences	Study of Engineering and the Science of War
From ½ past 9 to 10	*Tattoo at half past 9 o'clock—Roll call immediately after tattoo Signal to extinguish lights, and inspection of rooms at 10 o'clock*				

Source: U.S. Military Academy Staff Records, vol. 6, 1854–59, 198–99, USMA Spec Coll.

APPENDIX OF TABLES

TABLE 4. Officers performing staff functions, 1835

Number of commissioned officers	603
Number of officers allocated to general staff bureaus	94 Officers on general staff—27 1 Adjutant general 2 Inspector generals 1 Quartermaster general 4 Quartermaster general assistants 1 Commissary of subsistence 1 Surgeon general 1 Paymaster general 14 Paymaster general Assistants 2 Storekeepers Surgeons—12 Assistant surgeons—55
Number of line officers seconded to General Staff Bureaus from infantry, dragoon, and artillery regiments (not including officers assigned as assistant quartermasters within line regiments)	128 23 on recruiting duty 16 on ordnance duty 24 on Topog engineer duty 21 on engineer duty 25 on instructor duty at USMA 19 on special staff duty
Number of regular army officers performing staff functions	155 (25.7%) not including surgeons
Number of army posts	Eastern Department—29 Western Department—21 Arsenals/Depots—13

Source: Adjutant General's Office, *Army Register* (Washington: C. Alexander, 1835).

APPENDIX OF TABLES

TABLE 5. Officers employed in staff functions during the antebellum period

Year	Officers employed in general staff bureaus or seconded to general staff duties from the line	Military Posts
1835	222 out of 603 officers (36.8%)	Eastern Department, 29 Western Department, 21 Arsenals, 13
1840	250 out of 735 officers (34%)	Eastern Department, 37 Western Department, 20 Arsenals, 18
1845	233 out of 733 officers (31.7%)	Eastern Department, 36 Western Department, 21 Arsenals, 17
1850	276 out of 884 officers (31%)	Eastern Department, 35 Western Department, 53 Arsenals, 18
1855	344 out of 948 officers (27.5%)	Eastern Department, 22 Western Department, 49 Arsenals, 24
1860	242 out of 1,083 officers (22%)	Eastern Department, 12 Western Department, 67 Arsenals, 23

Source: Adjutant General's Office, *Army Registers* (Washington: C. Alexander, 1835–60).

TABLE 6. Actual and possible membership of the Napoleon Club

Actual Members of the Napoleon Club in the Antebellum Period					
West Point Number	Instructor Rank & Name	Branch	Years	Civil War Rank/Appointment
361	Lt. (Rtd) Dennis H. Mahan	Engineer	1824–26 1830–71	Professor, President of the Napoleon Club
709	Capt. George W. Cullum	Engineer	1847–51 1852–55	Bvt. Maj Gen U.S. Army (Departmental and Army Chief Engineer)
1028	1st Lt. George H Thomas	Artillery	1851–54	Maj. Gen U.S. Army (Commander Army of the Cumberland)
1077	1st Lt. Samuel Jones	Artillery	1846–51	Col. CSA Inf. killed 1862
1117	2nd Lt. Barton S. Alexander	Engineer	1848–52	Bvt. Brig. Gen. U.S. Army (Corps Chief Engineer, Engineer Board)
1118	1st Lt. Gustavus W. Smith	Engineer	1849–54	Maj. Gen. CSA (Corps Commander)
1167	2nd Lt. William B. Franklin	Topog	1848–51	Bvt. Maj Gen. U.S. Army (Corps Commander)
1176	2nd Lt. Joseph J. Reynolds	Artillery	1846–55	Bvt. Maj. Gen. U.S. Army (Corps Commander)
1178	1st Lt. Henry F. Clarke	Artillery	1848–51	Bvt. Maj. Gen. U.S. Army (Departmental and Army Chief Commissary Officer)
1234	1st Lt. William F. Smith	Topog	1846–48 1855–56	Bvt. Maj. Gen. U.S. Army (Corps Commander)
1238	1st Lt. Fitz John Porter	Artillery	1849–55	Bvt. Maj. Gen. U.S. Army (Corps Commander)
1241	1st Lt. Henry Coppée	Artillery	1848–49 1850–55	Civilian

APPENDIX OF TABLES

1255	1st Lt. Edmund K. Smith	Infantry	1849–52	Lt. Gen. CSA (Departmental Commander)
1273	2nd Lt. George B. McClellan	Engineer	1848–51	Maj. Gen. U.S. Army (Army Commander)
1275	1st Lt. John G. Foster	Engineer	1855–57	Bvt. Maj. Gen. U.S. Army (Departmental Commander)
1325	1st Lt. Cadmus Wilcox	Infantry	1852–57	Maj. Gen. CSA (Division Commander)
1350	1st Lt. John Gibbon	Artillery	1854–59	Bvt. Maj. Gen. U.S. Army (Corps Commander)
1357	1st Lt. Thomas H. Neill	Infantry	1853–57	Bvt. Maj. Gen. U.S. Army (Division Commander)
1407	2nd Lt. Quincy A. Gilmore	Engineer	1852–56	Bvt. Maj. Gen. U.S. Army (Corps Commander)
1415	1st Lt. Absalom Baird	Artillery	1853–59	Bvt. Maj. Gen. U.S. Army (Division Commander)
1465	2nd Lt. John A. Mebane	Artillery	1851–54	Died 1854
1579	Bvt. 2nd Lt. James B. McPherson	Engineer	1853–54	Brig. Gen. U.S. Army (Army Commander)
Possible Members of the Napoleon Club in the Antebellum Period				
542	Lt. Col. Robert E. Lee	Engineer	1852–55	Lt. Gen. CSA (Army Commander)
1206	1st Lt. William G. Peck	Topog	1847–55	Civilian
1232	2nd Lt. E. B. Hunt	Engineer	1846–49	Major U.S. Army (Departmental Chief Engineer, Experiments Officer) killed 1863
1297	2nd Lt. Richard H. Rush	Artillery	1846–47	Colonel U.S. Volunteers (Brigade Commander)

APPENDIX OF TABLES

1331	Capt. John C. Symmes	Artillery Ordnance	1847–49 1855–56	Retired Disabled
1259	1st Lt. J. M. Hawes	Dragoon	1848–50	Brig. Gen. CSA (Cavalry Brigade Commander)
1374	2nd Lt. Rufus A. Roys	Engineer	1848–50	Killed 1850
1279	2nd Lt. Jesse L. Reno	Ordnance	1849	Maj. Gen. U.S. Army (Corps Commander) killed 1862
1369	2nd Lt. William Trowbridge	Engineer	1848–51	Civilian Engineer assisting U.S. Army 1861–65
1272	2nd Lt. C. Seaforth Stewart	Engineer	1849–54	Bvt. Col. U.S. Army (Division and District Chief Engineer)
1336	1st Lt. Daniel Van Buren	Artillery	1849–50	Bvt. Brig. Gen. U.S. Army (Corps and Departmental Chief of Staff)
1377	2nd Lt. Joseph C. Clark	Artillery	1849–51	Maj. U.S. Army (Artillery Battalion Commander) wounded 1862
1165	2nd Lt. John Abert	Infantry	1848–50	Maj. U.S. Army (Corps Topog Officer) wounded 1862
1410	2nd Lt. Thomas J. Haines	Artillery	1850 1852–53	Bvt. Brig. Gen. U.S. Army (Departmental Chief Commissary Officer)
1292	2nd Lt. Charles C. Gilbert	Infantry	1850–55	Bvt. Col. U.S. Army (Division Commander, District Provost Marshall and Mustering Officer)
1283	1st Lt. Edward C. Boyton	Artillery	1848–55	Bvt. Major U.S. Army (Adjutant at USMA)

APPENDIX OF TABLES

1266	1st Lt. Henry B. Clitz	Infantry	1848–55	Bvt. Brig. Gen. U.S. Army (Regiment Commander) wounded 1862
1290	1st Lt. Truman Seymore	Artillery	1850–53	Bvt. Maj. Gen. U.S. Army (Division Commander)
1133	1st Lt. Seth Williams	Artillery	1850–53	Bvt. Maj. Gen U.S. Army (Army Inspector General)
1413	1st Lt. Beekman du Barry	Artillery	1853–54 1859–61	Bvt. Brig. Gen. U.S. Army (Departmental Chief Commissary Officer)
1414	1st. Lt. Delvan D. Perkins	Artillery	1850–56	Maj. U.S. Army (Corps and Bureau Asst. Adj. Gen.)
1417	2nd Lt. Milton Cogswell	Infantry	1851–56	Bvt. Col. U.S. Army (Regiment Commander and Corps Provost Marshall)
1262	1st Lt. Delos B. Sacket	Dragoons	1850–55	Bvt. Maj. Gen U.S. Army (Corps/Army Inspector General)
1370	2nd Lt. Andrew J. Donelson	Engineer	1852–53 1855–56	Died 1859
1312	1st Lt. David R. Jones	Infantry	1851–53	Officer CSA (rank unknown) killed 1863
1420	2nd Lt. Chauncey McKeever	Artillery	1851–55	Bvt. Brig. Gen. U.S. Army (Corps Chief of Staff, Departmental Asst. Adj. Gen., Bureau Asst. Provost Marshall)
1371	1st Lt. James C. Duane	Engineers	1852–54 1858–61	Bvt. Brig. Gen. U.S. Army (Army Chief Engineer)
1418	2nd Lt. Edward D. Stockton	Infantry	1852–56	Died 1857

APPENDIX OF TABLES

1085	Capt. Robert S. Garnett	Artillery	1852–54	Brig. Gen. CSA killed 1861
1455	2nd Lt. Joseph H. Wheelock	Artillery	1851–55	Col. U.S. Volunteers died 1862
1506	2nd Lt. Alexander J. Perry	Artillery	1852–58	Bvt. Brig. Gen. U.S. Army (Corps Quartermaster)
1537	2nd Lt. Newton F. Alexander	Engineer	1852–53	Died 1858
1527	2nd Lt. Roger Jones	Mounted Riflemen	1852–54	Maj. U.S. Army (Division and Departmental Inspector General)
1419	1st Lt. Edward R. Platt	Artillery	1855–59	Maj. U.S. Army (Army Judge Advocate General)
1101	1st Lt. Anderson D. Nelson	Infantry	1853–55	Bvt. Col. U.S. Army (Corps and Departmental Inspector General)
1498	2nd Lt. Alexander Piper	Artillery	1853–54	Bvt. Lt. Col. U.S. Army (Corps Chief Artillery Officer)
1344	2nd Lt. James B. Fry	Artillery	1853–59	Bvt. Maj. Gen. U.S. Army (Provost Marshal General)
1499	2nd Lt. James Thompson	Artillery	1854–57	Bvt. Lt. Col. U.S. Army (Artillery Battalion Commander)
936	Capt. William H. T. Walker	Infantry	1854–56	Maj. Gen. CSA (Division Commander) killed 1864
1337	1st Lt. Samuel F. Chalfin	Artillery	1854–59	Bvt. Col. U.S. Army (Bureau Asst. Adj. Gen.)
1536	2nd Lt. Thomas Lincoln Casey	Engineers	1854–59	Bvt. Lt. Col. U.S. Army (District and Bureau Asst. Engineer)
1412	1st Lt. William Silvey	Artillery	1854–57	Bvt. Lt. Col. U.S. Army (District Asst. Adj. Gen.)

1581	2nd Lt. Joshua W. Sill	Ordnance	1854–57	Brig. Gen. U.S. Volunteers (Brigade Commander) killed 1862
1467	2nd Lt. Robert Ransom	Dragoons	1854–55	Maj. Gen. CSA (Cavalry Brigade and Departmental Commander)
708	Capt. John G. Barnard	Engineer	1855–56	Maj. Gen. U.S. Army (Army Chief Engineer)
1439	2nd Lt. Thorton A. Washington	Infantry	1855–56	Maj. CSA (State Asst. Adj. Gen.)
1461	1st Lt. Adam J. Slemmer	Artillery	1855–59	Bvt. Brig. Gen. U.S. Army (Regiment Commander) wounded 1862
1421	2nd Lt. William H. Lewis	Infantry	1855–56	Bvt. Lt. Col. U.S. Army (Chief Commissary Officer, Indian Territory and New Mexico)
1585	2nd Lt. John M. Schofield	Artillery	1855–60	Bvt. Maj. Gen. U.S. Army (Corps Commander)
1552	2nd Lt. George B. Cosby	Mounted Riflemen	1855–57	Brig. Gen. CSA (Army Chief of Staff and Cavalry Brigade Commander)
1519	1st Lt. John C. Kelton	Infantry	1858–59 1860–61	Bvt. Col. U.S. Army (Brigade Commander, Army/Bureau Asst. Adj. Gen.)
1084	Lt. Col John F. Reynolds	Artillery	1860–61	Maj. Gen. U.S. Volunteers (Corps Commander) killed 1863
966	Capt. William J. Hardee	Dragoons	1856–60	Lt. Gen. CSA (Corps and Departmental Commander)

APPENDIX OF TABLES

1121	Capt. James G. Benton	Ordnance	1857–61	Bvt. Col. U.S. Army (Bureau Senior Staff, Ordnance Board)
1583	2nd Lt. Francis J. Shunk	Ordnance	1855–57	Bvt. Maj. U.S. Army (Departmental Staff, Ordnance)
1495	2nd Lt. James St. C. Morton	Engineer	1855–57	Bvt. Brig. Gen. U.S. Army (Army Chief Engineer) killed 1864
1712	2nd Lt. David C. Houston	Engineer	1856–57	Bvt. Col. U.S. Army (Departmental Chief Engineer)
1651	2nd Lt. John T. Greble	Artillery	1856–60	1st Lt U.S. Army (Artillery) killed 1861
1267	1st Lt. William H. Wood	Infantry	1855–56	Lt. Col. U.S. Army (Regimental Commander, Division Provost Marshall)
1433	1st Lt. Charles W. Fields	Dragoons	1856–61	Maj. Gen. CSA (Corps Commander)
1300	1st Lt. Orren Chapman	Dragoons	1855–56	Died 1859
1512	2nd Lt. Robert Williams	Dragoons	1854 1857–61	Bvt. Brig. Gen. U.S. Army (Bureau Asst. Adj. Gen. Officer)
1381	1st Lt. Benjamin D. Forsythe	Infantry	1857	Died 1861
1590	1st Lt. Henry C. Symonds	Artillery	1857–61	Bvt. Col. U.S. Army (District Commissariat Officer)
1762	2nd Lt. E. Porter Alexander	Engineer	1858–60	Brig. Gen. CSA (Corps and Army Chief Artillery Officer)

APPENDIX OF TABLES

1640	2nd Lt. John Pegram	Dragoons	1857	Col. CSA (Chief Army Engineer Officer, Cavalry Regiment Commander)
1689	2nd Lt. Alexander S. Webb	Artillery	1857–61	Bvt. Maj. Gen. U.S. Army (Brigade/Division Commander)
1645	1st Lt. John R. Smead	Artillery	1857–59	Capt. U.S. Army (Artillery Battery Commander) killed 1862
1761	2nd Lt. Richard K. Meade	Engineer	1857–59	Capt. CSA (Corps Ordnance and Engineer Officer) killed 1862
1566	1st Lt. Henry Douglass	Infantry	1858–61	Bvt. Maj. U.S. Army (Battalion Command)
1565	1st Lt. Alexander McD. McCook	Infantry	1858–61	Bvt. Maj. Gen. U.S. Army (Corps and District Command)
1387	1st Lt. Richard L. Dodge	Infantry	1858–60	Bvt. Lt. Col. U.S. Army (State Chief Mustering/Disbursing Officer)
1752	1st Lt. James McMillan	Infantry	1858–61	Bvt. Maj. U.S. Army (Bureau Asst. Mustering/Disbursing Officer, Provost Office)
1775	2nd Lt. Francis Beach	Artillery	1858–59	Bvt. Lt. Col. U.S. Army (Brigade Commander)
1798	2nd Lt. William C. Paine	Engineer	1858–59	Capt. U.S. Army (Departmental Chief Engineer) Disabled 1863
1711	2nd Lt. George W. Snyder	Engineer	1859–60	Bvt. Maj. U.S. Army (Division Chief Engineer) died 1861
1500	1st Lt. Caleb Huse	Artillery	1852–59	Major CSA (Purchasing Agent)

APPENDIX OF TABLES

1424	2nd Lt. Rufus Saxton	Artillery	1859–60	Bvt. Brig. Gen. U.S. Army (Division/District Chief Quartermaster)
1451	1st Lt. Gouverneur K. Warren	Engineer	1859–61	Bvt. Maj. Gen. U.S. Army (Army Chief Engineer, Corps Commander)
1826	2nd Lt. Samuel D. Lockett	Engineer	1859–61	Col. CSA (Field Engineer)
1437	2nd Lt. Samuel B. Holabird	Infantry	1859–61	Bvt. Brig. Gen. U.S. Army (Departmental Chief Quartermaster)
1636	2nd Lt. Charles N. Turnbull	Topog	1859–60	Bvt. Col. U.S. Army (Engineer Battalion Commander, Chief Corps and Departmental Engineer)
1538	1st Lt. George H. Mendell	Engineer	1859–63	Bvt. Col. U.S. Army (Engineer Battalion Commander)
1745	2nd Lt. Herman Biggs	Infantry	1859–61	Bvt. Brig. Gen. U.S. Army (Departmental Chief Quartermaster)
1677	2nd Lt. Cyrus B. Comstock	Engineer	1859–61	Bvt. Brig. Gen. U.S. Army (Army Chief Engineer)
1580	1st Lt. William P. Craighill	Engineer	1859–63	Bvt. Lt. Col. U.S. Army (Division Chief Engineer)
1681	2nd Lt. Junius B. Wheeler	Cavalry Topog	1859–61 1861–63	Bvt. Lt. Col. U.S. Army (Division Chief Engineer)
1718	1st Lt. Herbert A. Hascall	Artillery	1860–61 1863–67	Bvt. Lt. Col. U.S. Army (Brigade Quartermaster)
1589	1st Lt. Thomas M. Vincent	Artillery	1860–61	Bvt. Brig. Gen. U.S. Army (Senior Staff, Bureau Asst. Adj. Gen.)

APPENDIX OF TABLES

1601	1st Lt. Walworth Jenkins	Artillery	1860–61	Capt. U.S. Army (Division/District Chief Quartermaster)
1754	2nd Lt. Samuel S. Carrol	Infantry	1860–61	Bvt. Maj. Gen. U.S. Army (Brigade Commander)
1678	1st Lt. Godfrey Weitzel	Engineer	1860–61	Bvt. Maj. Gen. U.S. Army (Corps Commander)
1409	1st Lt. Stephen V. Benét	Ordnance	1860–64	Bvt. Lt. Col. U.S. Army (Bureau Staff, Ordnance Board)
1825	2nd Lt. William E. Merrill	Engineers	1860–61	Bvt. Col. U.S. Army (Army Chief engineer)
1683	2nd Lt Samuel Beck	Artillery	1860–61	Bvt. Brig. Gen U.S. Army (Corps/Bureau Asst. Adj. Gen.)
1832	2nd Lt. Edward G. Bush	Infantry	1860–61	Bvt. Maj. U.S. Army (Regimental Commander, Departmental Provost Marshall)
1353	1st Lt. Charles Griffin	Artillery	1860–61	Bvt. Maj. Gen. U.S. Army (Corps Commander)
1726	1st Lt. Wesley Owens	Cavalry	1860–61	Bvt. Lt. Col. U.S. Army (Regimental Commander, Inspector Provost Marshall Generals Department)

Sources: Mr. Bowman to Lincoln, May 1863, USMA Letters Received, 1840–90, USMA Spec. Coll. Cited in Griess, "Mahan," 237. Names of possible members taken from *Officers and Members of the West Point Army Mess, 1841 to 1880* (West Point: USMA Press, 1880). Details of individuals from George Cullum, *Biographical Register*; and Ellsworth Eliot, Jr., *West Point in the Confederacy* (New York: G. A. Baker, 1941); and Charles B. Hall, *Military Records of General Officers of the Confederacy* (New York: Steck, 1898); and Frederick Phisterer, *Statistical Record of the Armies of the United States* (New York: The Blue and the Gray Press, 1909).

APPENDIX OF TABLES

TABLE 7. Participants in Mahan's advanced engineer studies program

West Point Number	Rank (as an Instructor) Name	Branch	At West Point as Instructor	Civil War Rank (Command/ Branch)
587	Capt. Alexander J. Swift	Engineer	1841–46	Killed Mexico 1847
1059	Zealous B. Tower	Engineer	1842–43	Bvt. Maj. Gen U.S. Army (District Chief Engineer)
1060	Horatio H. Wright	Engineer	1842–44	Bvt. Maj. Gen U.S. Army (Corps Commander)
1115	William Rosecrans	Engineer	1843–47	Bvt. Maj. Gen. U.S. Army (Army Commander)
1112	John Newton	Engineer	1843–46	Bvt. Maj. Gen. U.S. Army (Corps Commander)
1232	2nd Lt. E.B. Hunt	Engineer	1846–49	Major U.S. Army (departmental Chief Engineer, Experiments Officer) killed 1863
1234	1st Lt. William F. Smith	Topog	1846–48 1855–56	Bvt. Maj. Gen. U.S. Army (Corps Commander)
1206	1st Lt. William G. Peck	Topog	1847–55	Civilian
1111	Henry L. Eustis	Engineer	1847–49	Brig. Gen. U.S. Volunteers (Regimental Commander)
709	Capt. George W. Cullum	Engineer	1847–51 1852–55	Bvt. Maj Gen U.S. Army (Departmental and Army Chief Engineer)

APPENDIX OF TABLES

1117	2nd Lt. Barton S. Alexander	Engineer	1848–52	Bvt. Brig. Gen. U.S. Army (Corps Chief Engineer, Engineer Board)
1167	2nd Lt. William B. Franklin	Topog	1848–51	Bvt. Maj Gen. U.S. Army (Corps Commander)
1374	2nd Lt. Rufus A. Roys	Engineer	1848–50	Killed 1850
1273	2nd Lt. George B. McClellan	Engineer	1848–51	Maj. Gen. U.S. Army (Army Commander)
1369	2nd Lt. William Trowbridge	Engineer	1848–51	Civilian Engineer assisting U.S. Army 1861–65
1272	2nd Lt. C. Seaforth Stewart	Engineer	1849–54	Bvt. Col. U.S. Army (Division and District Chief Engineer)
1118	1st Lt. Gustavus W. Smith	Engineer	1849–54	Maj. Gen. CSA (Corps Commander)
1370	2nd Lt. Andrew J. Donelson	Engineer	1852–53 1855–56	Died 1859
1371	1st Lt. James C. Duane	Engineers	1852–54 1858	Bvt. Brig. Gen. U.S. Army (Army Chief Engineer)
1537	2nd Lt. Newton F. Alexander	Engineer	1852–53	Died 1858
1407	2nd Lt. Quincy A. Gilmore	Engineer	1852–56	Bvt. Maj. Gen. U.S. Army (Corps Commander)
1579	Bvt. 2nd Lt. James B. McPherson	Engineer	1853–54	Brig. Gen. U.S. Army (Army Commander)
1536	2nd Lt. Thomas Lincoln Casey	Engineers	1854–59	Bvt. Lt. Col. U.S. Army (District and Bureau Asst. Engineer)

1494	George L. Andrews	Engineer	1854–55	Bvt. Maj. Gen. U.S. Volunteers (Departmental Chief of Staff, Corps Commander)
708	Capt. John G. Barnard	Engineer	1855–56	Maj. Gen. U.S. Army (Army Chief Engineer)
1275	1st Lt. John G. Foster	Engineer	1855–57	Bvt. Maj. Gen. U.S. Army (Departmental Commander)
1495	2nd Lt. James St. C. Morton	Engineer	1855–57	Bvt. Brig. Gen. U.S. Army (Army Chief Engineer) killed 1864
1712	2nd Lt. David C. Houston	Engineer	1856–57	Bvt. Col. U.S. Army (Departmental Chief Engineer)
1761	2nd Lt. Richard K. Meade	Engineer	1857–59	Capt. CSA (Corps Ordnance and Engineer Officer) killed 1862
1762	2nd Lt. E. Porter Alexander	Engineer	1858–60	Brig. Gen. CSA (Corps and Army Chief Artillery Officer)
1798	2nd Lt. William C. Paine	Engineer	1858–59	Capt. U.S. Army (Departmental Chief Engineer) disabled 1863
1711	2nd Lt. George W. Snyder	Engineer	1859–60	Bvt. Maj. U.S. Army (Division Chief Engineer) died 1861
1451	1st Lt. Gouverneur K. Warren	Engineer	1859–61	Bvt. Maj. Gen. U.S. Army (Army Chief Engineer, Corps Commander)
1826	2nd Lt. Samuel D. Lockett	Engineer	1859–61	Col. CSA (Field Engineer)

APPENDIX OF TABLES

1636	2nd Lt. Charles N. Turnbull	Topog	1859–60	Bvt. Col CSA. U.S. Army (Engineer Battalion Commander, Chief Corps/ Departmental Engineer)
1538	1st Lt. George H. Mendell	Engineer	1859–63	Bvt. Colonel U.S. Army (Engineer Battalion Commander)
1677	2nd Lt. Cyrus. B. Comstock	Engineer	1859–61	Bvt. Brig. Gen. U.S. Army (Army Chief Engineer)
1580	1st Lt. William P. Craighill	Engineer	1859–63	Bvt. Lt. Col. U.S. Army (Division Chief Engineer)
1681	2nd Lt. Junius B. Wheeler	Cavalry Topog	1859–61 1861–63	Bvt. Lt. Col. U.S. Army (Division Chief Engineer)
1678	1st Lt. Godfrey Weitzel	Engineer	1860–61	Bvt. Maj. Gen. U.S. Army (Corps Commander)
1825	2nd Lt. William E. Merrill	Engineer	1860–61	Bvt. Col. U.S. Army (Army Chief engineer)

Sources: Cullum's *Register*, together with the document *Officers and Members of the West Point Officers Mess, 1841 to 1880* (West Point: USMA Press, 1880) reveal the assistant instructors under Mahan during this period. For course details, see Joseph Totten to Richard Delafield, Washington, 9 January 1844, Adjutant General War Department, USMA, 1812–67, XI, 366–67; Lt. Z. Tower to Richard Delafield, 8 April 1843, Lt. G. W. Smith to Mahan, 2 May 1846; Lieutenants Newton, Rosecrans, and Smith to Superintendent Henry Brewerton, 12 May 1846; Mahan to Brewerton, 8 June 1846; and Rosecrans to Brewerton, 12 February 1847, all in USMA, Letters Received, 1840–90, USMA Spec. Coll.; see also Griess, "Mahan," 163–64.

APPENDIX OF TABLES

TABLE 8. West Point graduates during the Civil War and their prewar experience

Wartime Appointment	Of 1,135 USMA Graduates	Antebellum Experience		
		USMA Instructor	ENGINEER Duty	STAFF Duty
General officer appointments	368	129	105	94
Commanding generals	4	2	2	3
Field army commanders	28	9	10	14
Corps commanders	61	22	20	22
Division commanders	107	22	26	33
Brigade commanders	120	16	21	22
Military department command*	40	14	11	17
Military district command*	48	9	15	18
Captain-colonel staff appointments	797**	126	132	149
In general staff departments	92	24	28	41
In geographic department staff	155	20	31	44
In geographic district staff	211	26	30	39
In field army staff	264	50	40	7
Commanding depots/arsenals	57	6	3	18

*Often held simultaneous to a field command.

**Many graduates served in staff positions before rising to general officer appointments, and these are included in this total.

Note: It is difficult to categorize general officer status when having to deal with exigencies with brevet rank, differences between volunteer and regular ranks, and the paucity of information regarding Confederate general officers. In this we are greatly aided by *The Centennial of the United States Military Academy at West Point, New York, 1802–1902*, appendix 1; Eliot, *West Point in the Confederacy*; Hall, *Military Records of General Officers of the Confederacy*; and Phisterer, *Statistical Record of the Armies of the United States*.

NOTES

Introduction

1. President Lincoln to General Grant, telegram, 17 August 1864: "I have seen your dispatch expressing your unwillingness to break your hold where you are. Neither am I willing. Hold on with a bulldog grip, and chew and choke as much as possible." Accessed 12 December 2012 at www.archives.gov/exhibits/american_originals/petersbg.html.

2. Quotation from Jones, *Personal Reminiscences of General Robert E. Lee*, 40; from address by Jubal Early to Washington and Lee University, 19 January 1872.

3. General Sherman to Secretary of War E. M. Stanton, telegram, 12 March 1865, in Sherman, *Memoirs*, 296.

4. W. T. Sherman to his brother, December 1863, in Royster, *The Destructive War*, 322.

5. Jones, *Personal Reminiscences*, 43.

6. *Doctrine* is defined as that which is taught, and refers here specifically to those subjects taught at West Point; see *Shorter Oxford English Dictionary*, 5th ed., 1:727. Also see Merriam-Webster at http://www.merriam-webster.com/dictionary/doctrine.

7. For technology-based "military revolutions," see Roberts, *The Military Revolution, 1560–1660*; Parker, *The Military Revolution*; Rogers, *The Military Revolution Debate*; and Black, *A Military Revolution?*

8. Vauban, *Le Triomphe de la Méthode*. For contemporary references of Vauban's applicability to open warfare, see Marquis de Puységur, *Art de la guerre par principes et par règles*, 1:2, 75; Le Comte Turpin de Crissé, *Essai sur l'art de la guerre*, 2:136, 139; Guibert, *General Essay on Tactics*, xlvii; and a quotation of Napoleon in Chandler, *The Campaigns of Napoleon*, 135. See also Carl von Clausewitz, *On War*, 153.

9. For contemporary definition of science in the United States, see Woodward, *A System of Universal Science*, 10–11; for "military science," see 331–38.

10. For discussion of the place of science in eighteenth-century schooling, see Hornberger, *Scientific Thought in the American Colleges, 1638–1800*. For branches

of military science and their relationship to the art of war, see Halleck, *Elements of Military Art and Science*, 37–38.

11. Winfield Scott continued to use the term *science of movements* after West Point adopted the word *strategy*: see Scott, *Infantry Tactics*, title page.

12. West Point provided instruction in these areas and branch tactics. Halleck named five elements of the art of war: fortifications, tactics, strategy, logistics, and what he called the polity of war. Halleck, *Elements of Military Art and Science*, 37. He saw these as elements of the military art and the branches of military science interchangeably. Subsequent West Point instructors called these same elements the scientific parts or branches to the art of war. See Wheeler, *A Course of Instruction in the Elements of the Art and Science of War for the Use of Cadets of the United States Military Academy*.

13. Mahan, *Advanced-Guard, Outpost, and Detachment Service of Troops*, 1863, 169.

14. This distinction between strategy and tactics was in Heinrich von Bülow, *The Spirit of Modern System of War*, 86–87. It was used at West Point thereafter. See Halleck, *Elements of Military Art and Science*, 38, and Mahan, "Composition of Armies," lithographic notes, 1841 FT 355 (2814), U.S. Military Academy, Special Collection (hereafter USMA Spec. Coll.), 2.

15. Jomini, *An Exposition of First Principles of Grand Military Combinations and Movements*, 3.

16. Pascal, *Pensées* and *Scientific Treatises*, 33:432–34. For the pervasiveness of such thinking in the military world, see Gat, *A History of Military Thought*, 37; and Guerlac, "Vauban: The Impact of Science on War."

17. The sublime: "Characterized by nobility and grandeur, impressive, exalted, raised above ordinary human qualities . . . essential qualities of great art": Harmon and Holman, *A Handbook to Literature*. This definition is drawn from the use of the word *sublime* in eighteenth-century literary sources—most notably Edmund Burke's *A Philosophical Inquiry into the Origin of Our Ideas of the Sublime and the Beautiful* (1756) and Kant's *Critique of Judgement* (1790). Both distinguish beauty and the sublime, the first being finite and easily appreciated, the second being infinite and powerful, invoking passion and release from rational thinking. *Coup d'œil* is French for "glance of the eye" and refers to the ability to assess the real and potential aspects of any situation by rapid observation, leading to quick decisions regarding the military options in that circumstance. See Turpin de Crissé, *Essai sur l'art de la guerre*, 1:5.

18. The importance of genius and science in war presents a dichotomy that is fundamental to understanding antebellum military affairs. The West Point academy inherited a French Enlightenment emphasis on science. This contrasts with a contemporary Romantic view held in England and Prussia that genius trumped all. The evolution of these perspectives is relatively easy to follow. For the Enlightenment view, see how the Marquis de Puységur favored science over the sublime in *Art de la guerre par principes et par règles*, 1:iii, 2, and 75. Le Comte Turpin de Crissé felt that genius alone was not enough; it took study of military science to perfect innate tal-

ents. See his *Essai sur l'art de la guerre*, 2:4–6. Jacques Antoine Hippolyte de Guibert championed science in his *General Essay on Tactics*, xlv–xlvii, liii. Heinrich von Bülow carried this forward in *The Spirit of the Modern System of War*. Baron de Jomini acknowledged both genius and science, but favored the latter in understanding strategy: *An Exposition of the First Principles of Grand Military Combinations and Movements*, and Jomini, *Treatise on Grand Military Operations*. West Point instruction was dominated by the Enlightenment view: see Gay de Vernon, *Treatise on the Science of War and Fortification*; Mahan, "Composition of Armies"; and Halleck, *Elements of Military Art and Science*. For the Anglo-German view of genius, see Maurice de Saxe, *My Reveries upon the Art of War*, 190–92, 259–97. Saxe influenced Henry Lloyd; see *The History of the Late War in Germany*, 2:vi. C. Malorti de Martemont criticizes the very notion of a science to war in his introduction to the first English translation of von Bülow, *The Spirit of the Modern System of War*, ii–iv. The most comprehensive rendering of the importance of genius is in Clausewitz, *On War*, chapter 3.

19. Isaiah Berlin believed that an affinity for Romanticism has come to prevail over late twentieth-century interpretations of the past. See P. Watson, *Ideas: A History from Fire to Freud*, 607; and Berlin, *Concepts and Categories*, 153–54. For the Military Enlightenment, see Lynn, *Battle: A History of Combat and Culture*, 190–217; Gat, *A History of Military Thought*, 56–107, 141–91; and Starky, *War in the Age of the Enlightenment, 1700–1789*, 33–68. For postmodern interpretations of nineteenth-century mathematics (and other sciences), see Dauben, "Mathematics," 152–53.

20. Clausewitz, *On War*, vol. 1, chapters 1 and 3.

21. For the narrow technical critique of the West Point curriculum, see Huntington, *The Soldier and the State: The Theory and Politics of Civil-Military Relations*, 195–211; Moten, *The Delafield Commission and the American Military Profession*, 35–36; Morrison, *"The Best School": West Point, 1833–1866*, 87–101, 153; and Morrison, "Educating the Civil War Generals: West Point, 1833–1861." For Jomimian bias of the curriculum, see Hittle, *Jomini and His Summary of the Art of War*, ii; Weigley, *Towards an American Army: Military Thought from Washington to Marshall*; and Weigley, *The American Way of War: A History of United States Military Strategy and Policy*. For rebuttal of Jominian influence, see Reardon, *With a Sword in One Hand and Jomini in the Other: The Problem of Military Thought in the Civil War North*.

22. Weigley, *The American Way of War*; Linn and Weigley, "The American Way of War Revisited."

23. Linn, *The Echo of Battle*, 3–4.

24. Huntington, *The Soldier and the State*, 59–79. For Levy-Buhl and Bloch, see Skinner, ed., *The Return of Grand Theory in the Social Sciences*, 181; and Marc Bloch, *The Historian's Craft*. Inherent in all these terms are intellectual "structures," patterns of thought with boundaries that are the basis for organizational culture and doctrine.

25. Kuhn, *Structure of Scientific Revolutions*, 181–204.

26. Referred to simply as "coastal fortifications" in government policy between 1820 and 1850, the policy became known as the "Third System" (two having previ-

ously been attempted in the 1790s and the early 1800s). See Totten, *Report of General J. G. Totten, Chief Engineer, on the Subject of National Defences*; M. Smith, "The Corps of Engineers and National Defense in Antebellum America, 1815–1860," note on page 2.

27. While other army institutions were established during the antebellum period to instruct specific branch tactics, none lasted more than a few years before disappearing for lack of congressional support. The academy alone continued to instruct military art and science throughout the era, and it was academy graduates (mostly faculty) who produced the first American treatises on military affairs. All definitive texts agree on this point: see Coffman, *The Old Army*; Morrison, *"The Best School"*; Skelton, *An American Profession of Arms*, chapter 13; Ambrose, *Duty, Honor, Country*; Pappas, *To the Point*; and Dupuy, *Where They Have Trod*.

28. See Hope, "A Scientific Way of War: Antebellum Military Science, West Point, and the Origins of American Military Thought," appendixes D and E. Accessed at http://qspace.library.queensu.ca/handle/1974/7326.

29. *La petite guerre* or "little war" was the name used to describe irregular warfare in mountainous or wooded terrain, involving small numbers of irregular troops operating in outposts or mobile bands, or by regular troops on detached service from a main army. See Grenier, *The First Way of War*.

30. Huntington, *The Soldier and the State*, 28–30.

31. Mahan, *Advanced-Guard*, 33.

1. Colonial and Early National Military Science

1. Walker, *Engineers of Independence*, 367n1.

2. Edward Winslow, "A Letter Sent from New England," written in the winter 1621–22, found on 12 January 2009 at http://www.nationalcenter.org/Pilgrims.html. He also described the muster of every man and boy with weaponry upon sight of a French ship.

3. Edward Johnson in Ferling, *A Wilderness of Miseries*, 10.

4. Vattel, *The Law of Nations*, 293; see also Lepore, *Name of War*, xvii, for discussion on the uses of the word *war* for military action taken against the Native Americans.

5. For a complete bibliography of American military imprints, see Riling, *The Art and Science of War in America*.

6. For a comprehensive look at early American reliance on classical literature, see Dederer, *War in America to 1775: Before Yankee Doodle*.

7. Riling, *The Art and Science of War in America*; Bland, *A Treatise of Military Discipline*; J. Muller, *A Treatise Containing the Elementary Part of Fortification*; Robins, *New Principles of Gunnery*; Stevenson, *Military Instructions for Officers Detached in the Field: Containing, a Scheme for Forming a Corps of a Partisan*.

8. For an enduring work on early colonial warfare, see Ferling, *A Wilderness of Miseries*; for the rudimentary nature of militia training, see Ahearn, *Rhetoric of War Training Day, the Militia, and the Military Sermon*.

9. Massachusetts surveyor and militia artillery captain Richard Gridley had for decades assisted Royal engineers in the colony and earned some fame in his successful participation of the siege and capture of Louisbourg in 1745. Age precluded him from great employment during the Revolution.

10. Whisker, *The Rise and Decline of the American Militia System*, 12–13; see also Boorstin, *The Americans: The Colonial Experience*, 356.

11. Ahearn, *Rhetoric of War Training Day*, 13, 145n9.

12. Chorley, *Armies and the Art of Revolution*, 164. For state of colonial militia in 1776, see Royster, *A Revolutionary People at War*.

13. This in fact had already become the status of training days in many of the colonies. See Mook, "Training Day in New England," and Ahearn, *Rhetoric of War Training Day*. For a detailed examination of militia service in the American colonies and young republic, see Whisker, *The Rise and Decline of the American Militia System*, 110.

14. Grenier examines the lessons that colonists learned from English experience in Ireland involving *guerre des postes*, the "feedfight," and the killing of noncombatants and forced removals in *The First Way of War*, 36–42, 103–4. This is substantiated in other works. See Canny "The Ideology of English Colonization: From Ireland to America," and Muldoon, "The Indian as Irishman." For American frontiersman identity, see Grenier, *The First Way of War*, 222–23.

15. Shy, "American Strategy: Charles Lee and the Radical Alternative."

16. For a superb account of the role and effectiveness of the militia during the revolution, see Kwasny, *Washington's Partisan War*.

17. Walker, *Engineers of Independence*, 5–6.

18. Walker, *Engineers of Independence*, 204.

19. Walker, *Engineers of Independence*, 23.

20. Baron von Steuben's manual was replaced in the 1820s by Winfield Scott's regulation handbook. Lockhart, *The Drillmaster of Valley Forge*, 186–96, 200, 301.

21. Kite, *Brigadier-General Louis Lebegue Duportail*, 47–50.

22. Walker, *Engineers of Independence*, 43, 49.

23. Risch, *Supplying Washington's Army*, 6–14. See also Saxe, *My Reveries upon the Art of War*, and Turpin de Crissé, *Essai sur l'art de la guerre*.

24. Quartermaster's Department, *A Sketch of the Organization of the Quartermasters' Department from 1774 to 1876*, 4–10.

25. See Walker, *Engineers of Independence*, 349–53, for a complete reprint of Duportail's report.

26. Kite, *Duportail*, 268.

27. Letter by Pierre L'Enfant dated 15 December 1784 in *Papers of the Continental Congress*, roll 98. Reproduced in Walker, *Engineers of Independence*, 354–63.

28. Skelton, *An American Profession of Arms*, 93.

29. Washington's "Sentiments" are reprinted in Millis, *American Military Thought*, 27.

30. One regiment was earmarked for the northeast frontier (the Canadas and New Brunswick). A second regiment would defend the Great Lakes Oswego-Mackinac frontier. A third would defend along the Ohio River, and the fourth would defend the Georgia-Carolina frontier. See Kohn, *Eagle and Sword*, 45–46.

31. Millis, *American Military Thought*, 24 and 26. Also, Fitzpatrick, ed., *The Writings of George Washington*, 26:374–98.

32. Millis, 26-27.

33. Millis, 26.

34. Kohn, *Eagle and Sword*, 50–70.

35. Kohn, *Eagle and Sword*, 86–87.

36. Kohn, *Eagle and Sword*, 81.

37. John Knox, "A Plan for the General Arrangement of the Militia of the United States," 1786, submitted to Senate 21 January 1790 as "Organization of the Militia." See *American State Papers—Military Affairs*, vol. 1, no. 2, p. 7 (hereafter referred to as *ASP MA* or *ASP Miscellaneous*); Kohn, *Eagle and Sword*, 129.

38. Kohn, *Eagle and Sword*, 131.

39. *Militia Act of 1792*, 2nd Cong., 1st sess., chap. 28, accessed at www.constitution.org/mil/mil_act_1792.htm. The Whiskey Rebellion was a brief uprising in western Pennsylvania by settlers disgruntled by a whiskey excise tax. It collapsed after President Washington federalized state militia and began marching on the insurrectionists. The St. Clair disaster occurred in November 1791 when Governor St. Clair of the Northwest Territories commissioned a punitive expedition against the Shawnee and Miami native bands, but it was decisively defeated on the Wabash River, resulting in over six hundred white dead and two hundred wounded. These Native Indians were subsequently defeated by an expedition led by Gen. Anthony Wayne in 1794.

40. *ASP MA*, vol. 1, 3rd Cong., 1st sess., no. 13, 28 February 1794, "Fortifications," 61–64; no. 14, 5 March 1794, "Arsenals and Amories," 65–66; no. 22, December 19, 1794, "Fortifications," 71–107; no. 24, January 28, 1795, "Fortifications," 107–8.

41. *ASP MA*, vol. 1, no. 13, 28 February 1794, "Fortifications," 61. See also vol. 1, no. 27 (25 March 1796), 172, for organization of the military establishment.

42. Duke de la Rochefoucault-Liancourt, *Travels through the United States of North America*, 2:625. The problems of oversight, appropriations, and provision of material and garrison troops are well documented in correspondence and official reports to the House and Senate committees of defense in Cooling, ed., *New American State Papers: Military Affairs*, 44–67 (hereafter referred to as *NASP MA*). For an American appraisal of the system, see Ellicott, *The Journal of Andrew Ellicott*, 282–83.

43. Of course, such a system led to mixed results. Of the states that did execute this alternative, New York appears to have benefited the most, spending $891,129 on city defenses. See Moore, *The Fortifications Board, 1816–1828, and the Definition of National Security*, chapter 1.

44. Rochefontaine's failure left a sour taste for the idea of such a school; see Ellicott, *Journal*, 282–83. Hamilton's initiative was later codified by McHenry in sub-

mission to John Adams, Article no. 39, "Military Academy and Reorganization of the Army," Communicated to Congress January 14th, 1800, in *ASP MA* 1:133–35.

45. *Annals of Congress*, 5th Cong., 15 May 1797 to 3 March 1799, 1419–21.

46. John Adams, Article no. 39, "Military Academy and Reorganization of the Army" Communicated to Congress January 14th, 1800, in *ASP MA*, 1:133–35.

47. James McHenry, Article no. 40, "Military Academy," communicated to Congress February 13th, 1800, in *ASP MA*, 1:142–44.

48. *ASP MA*, 1:229, 483. Quotations from *NASP MA*, vol. 16, no. 155, "Report of Committee of Military Affairs, Statement of History and Importance of the Military Academy, and Why It Should Not Be Abolished, 17 May 1834," 304. See also Kohn, *Eagle and Sword*, 193–95, 286–302.

49. Swift, *Memoirs*, 36.

50. This is a central theme in Crackel, *Mr. Jefferson's Army: Political and Social Reform of the Military Establishment*.

51. Pappas, *To the Point*, 25.

52. Pappas, *To the Point*, 45–59; see also Molloy, "Technical Education and the Young Republic: West Point as America's École Polytechnique, 1802–1833."

53. Noted men of science such as Robert Fulton and Eli Whitney were also members; see Wade, "A Military Offspring of the American Philosophical Society." Wade cites John Adams, Jefferson, Madison, and future presidents Monroe and John Quincy Adams, as well as a vice president, chief justice, New York mayor, two secretaries of war, two secretaries of the navy, eight senators, and five governors among the membership.

54. Skelton, *An American Profession of Arms*, 102–3.

55. Extracts from the Minutes of the U.S. Military Philosophical Society, at an occasional meeting, held Washington, 30 January 1808, USMA Spec. Coll.; Lt. Col. Jonathan Williams to President Thomas Jefferson, 1803, in USMA Spec. Coll.

56. Extracts from the Minutes of the U.S. Military Philosophical Society, at an occasional meeting, held in Washington, 30 January 1808, USMA Spec. Coll.; Wade, "A Military Offspring."

57. Williams to Jefferson, date unspecified, 1803, in USMA Spec. Coll.

58. The importance of a general education in mathematics for all officers is evidenced in the proposal that four of the eight professors for the "fundamental school" should teach mathematics. For general discussion on the enduring power of Enlightenment ideas, see Curti, *The Growth of American Thought*, 149–51, 155–57, and 166–67.

59. Williams to Decius Wadeworth, 13 August 1802, Jonathan Williams Papers, Lilly Library, Indiana University.

60. *Regulations of the United States Military Academy*, 1810, 2, USMC Spec. Coll.

61. *ASP MA*, vol. 1, no. 59, 191.

62. Peters, ed., *Public Statutes at Large of the United States of America*, 2:720–21. Emory Upton called this the most important decision regarding West Point after that establishing it in 1802. See Upton, *Military Policy of the United States* (1911), 91–93.

63. Swift, *Memoirs*, 20.

64. Duane, *American Military Library*, 1:ii.

65. Turpin de Crissé,, *Essai sur l'art de la guerre*, 2:131–39; Lloyd, *History of the Late War in Germany*, 2:133–49.

66. Duane, *American Military Library*, 1:133.

67. Duane, *American Military Library*, 1:140.

68. Duane, *American Military Library*, 1:141.

69. Bülow, *Spirit of the Modern System of War*, 92; Duane, *Military Library*, 1:82–85.

70. Duane, *American Military Library*, 1:86–87. Duane was careful not to use the word *strategy* in his reiteration, preferring instead to use the word *operation* to define the campaign plan that he was describing.

71. Bülow had embraced the Enlightenment commitment to human progress, once stating that "all productions of the human mind are successive." He therefore had little difficulty using theories and practices set down by Vauban and developing theories that had been started by *les Lumières* in the previous decades. Bülow, *Spirit of the Modern System of War*, 239.

72. Duane, *American Military Library*, 1:63.

73. Duane, *American Military Library*, 1:62–69; see also Bülow, *Spirit of the Modern System of War*, 183.

74. Bülow, *Spirit of the Modern System of War*, 23–24, 251; Duane, *American Military Library*, 1:65.

75. Bülow, *Spirit of the Modern System of War*, 183; Duane, *American Military Library*, 1:70–72, 78–82.

76. Hoyt, *Practical Instructions for Military Officers*, iii–vii; also see figures on 42 and 71.

77. Colson, *La Culture Strategique Americaine: L'influence de Jomini*, 28–29.

2. Army Reforms, 1815–1820

1. For American accounts of British activities in the Chesapeake, 1813, see *ASP MA*, 1:362–67; Coles, *The War of 1812*, 93; C. G. Muller, *The Darkest Days: The Washington Baltimore Campaign*, 13–41; Leckie, *The Wars of America*, 254–57; George, *Terror on the Chesapeake*; and Latimer, *1812: War with America*. British accounts of these activities are found in Horsman, *The War of 1812*, and Napier, *The Life and Correspondence of Admiral Charles Napier*.

2. For details of coast defenses in 1812–14, see Lewis, *Seacoast Fortifications*, 21–26; Kaufmann and Kaufmann, *Fortress America*, 207. For details of the defense of Craney Island, see Muller, *Darkest Days*, 34; Clary, *Fortress America*, 33–34; Hannings, *Forts of the United States*, 559. Also see reports to the Senate in *ASP MA*, 1:124, 377–78.

3. For pillage of Hampton, see Flanders, "Craney Island Battle Led to Burning of Hampton"; *Niles' Weekly Register*, 10 July 1813, 310–11.

4. The British were quick to blame the acts on a force of Chasseurs Britannique, French prisoners of war who had volunteered to serve the British in North Amer-

ica. A good firsthand account is also found in Rouse, ed., "The British Invasion of Hampton in 1813: The Reminiscences of James Jarvis." Many firsthand accounts and correspondence on this incident are found in *ASP MA*, 1:339–40, 362–67, 375–81. See also *Niles' Weekly Register*, Saturday, 24 July 1813, 333.

5. Dupuy, *Where They Have Trod*, 56–58; Cullum, *Biographical Register of the Officers and Graduates of the U.S. Military Academy at West Point, N.Y.*, 1:82.

6. "The depredations committed by the enemy during the last war along our coast by his occupying the mouths of our harbor, bays, and rivers, produced a universal sentiment . . . of the necessity of enlarging and strengthening the defences of our maritime frontier." Written in 1826, in *ASP MA*, vol. 3, paper no. 292, 185.

7. *ASP MA*, vol. 1, no. 133, 514, "Improvement and Increase in the Military Establishment, by James Monroe, 17 October 1814." Howe, *What Hath God Wrought*, 67.

8. The militia also proved to be more expensive than regulars, especially when factoring in postwar pensions. They were wasteful as well. During the War of 1812, more than 200,000 militia muskets were lost (worth $3,360,000) "mainly from that neglect and waste of public property which almost invariably attends the movements of newly-raised and inexperienced forces." Halleck, *Elements of Military Art and Science*, 150–51.

9. Upton, *Military Policy of the United States* (1911), 141. According to Francis Beirne, the costs of war (including pensions) were still being paid as late as 1940, *The War of 1812*, 391.

10. Wesley, *Guarding the Frontier*, 110. See also *Annals of Congress*, 15th Cong., 1st sess., 268, 273, 289–90, 293, 350, 1568–69, 1687–92.

11. Scott, *General Regulations for the Army; or, Military Institutions*. For a superb account of the influence of Jesup's bureau, see Wilson, *The Business of Civil War: Military Mobilization and the State*, 34–71.

12. For division of powers within the War Department, see Bernardo and Bacon, *American Military Policy*, 144, 151. The reduction of the regular army to 10,000 in March 1815 was in one respect a victory for Madison. Congress advocated reduction to 6,000, and it took Madison to mobilize Senate support to increase this to 10,000 in the legislative act. *Annals of Congress*, 13th Cong., 3rd sess., 244, 252, 286, 1196–1253. For Calhoun's statement, see pages 1215–16.

13. Joseph G. Swift to James Monroe, 21 March 1815, Washington, War Records, Engineer Department, Miscellaneous Letters Sent, I, 5–6, National Archives (hereafter NA).

14. Swift to Alexander Dallas, New York, 30 March 1815, and Alexander Dallas to Major Thayer, 20 April 1815, in the West Point Thayer Papers, 1808–72, section 2, 25/225, USMA Spec. Coll. (Also available at http://digital-library.usma.edu/libmedia/archives/thayer/thapap02.pdf, accessed 12 May 2009.)

15. Monroe to France's secretary Mardi de Marbois, Washington, 24 April 1815, West Point Thayer Papers, section 2, 31/225.

16. Seventh Annual Message to Congress on 15 December 1815, in J. W. Muller, ed., *Presidential Messages and State Papers*, 1:562–69. See also Howe, *What Hath God Wrought*, 80–81.

17. Price, "American Coastal Defense," 29, 60n20.

18. Simon Bernard was a graduate of the esteemed École Polytechnique and served as an engineer and aide-de-camp to Napoleon. He rose to the rank of brigadier and was present with Napoleon at Waterloo, an association which earned him great credibility in the United States and much resentment from American engineer officers who were bypassed when he received his appointment. See Smith, "The Corps of Engineers," 92–100.

19. Richardson, ed., *A Compilation of the Messages and Papers of the Presidents*, 2:576–77.

20. See *ASP Misc*, 7 January 1819, vol. 2, no. 462, 534.

21. Brig. Gen. Simon Bernard, Maj. Joseph Totten, and Capt. J. D. Elliot (U.S. Navy), "A Report for the Fortification of the Seacoasts," *ASP MA*, 2:304–13. This concept survived until the Civil War and was reiterated and modified in detail periodically. See *ASP MA*, vol. 3, no. 327, 283–302. In a document dated 25 February 1826, Chief Engineer Alexander Macomb delivered a complete plan including revised estimates for each fortifications site and garrison requirements for peace and for "siege" conditions.

22. *ASP MA*, vol. 7, 1819, no. 462, p. 534.

23. *ASP MA*, vol. 2, 310–12; Smith, "The Corps of Engineers," 42, 63n45.

24. Skelton, *An American Profession of Arms*, 243–48.

25. Smith, "The Corps of Engineers," 470. Here Smith outlines the yearly expenditures for fortifications and armaments between 1816 and 1861, with yearly averages being $894,318 and the total being $41,138,641.

26. For a synopsis of the influence and work of the Fortifications Board, see Moore, *The Fortifications Board, 1816–1828, and the Definition of National Security*. Other sources include Price, "American Coastal Defense"; Browning, *Two If by Sea*; Clary, *Fortress America*; S. Watson, "Knowledge, Interest, and Limits of Military Professionalism"; and Smith's "The Corps of Engineers." Primary source documents illustrative of the dominance of Fortifications Board work and internal improvement related to the Third System are found in *ASP MA*, vols. 1–3.

27. Watson, "Knowledge, Interest, and the Limits of Military Professionalism," 284–85.

28. For a description of how sentiment in Congress obstructed all attempts to retain a larger regular army or to create an efficient (standing) militia, see Ekirch, *The Civilian and the Military*, 63–65.

29. John C. Calhoun, "Reduction of the Army: Communicated to the House of Representatives, December 12, 1820," *ASP MA*, vol. 2, 188–93.

30. Calhoun was initially of the opinion that the purpose of the army was to garrison and help maintain fortifications. See *Annals of Congress*, 15th Cong., 1st sess., 22

December 1817, 496. He expanded his ideas and embraced the expansible army concept subsequently. Generals Scott and Gaines influenced Calhoun with this skeleton unit concept; see Scott to Calhoun, 20 August 1820, and Gaines to Calhoun, 27 July 1820, Letters Received, RG 107, NA. William Heath had originally proposed this expansible army idea to Washington in 1783. Washington included it in his "Sentiments on a Peace Establishment," and Hamilton later used it in debates in Congress. See Kohn, *Eagle and Sword*, 45–48. On Scott's influence, see Roger J. Spiller, "John C. Calhoun as Secretary of War, 1817–1825," 262–71; and T. Johnson, *Winfield Scott: Quest for Military Glory*, 85. Calhoun set a benchmark for the role of secretary of war as an advocate for the army. His centralized staff system remained essentially intact for the rest of the century. He changed the role of the regulars from being guardians of the frontiers to being a small body of professionals charged with preparing the nation for war. This required education in military science: "War has not only grown into an art—it is ennobled into a science." Quote from Macomb in Richards, *Memoir of Alexander Macomb*, ix.

31. Calhoun, "Reduction of the Army," 188.

32. Skelton states that when Congress cut private positions by 50 percent but only cut 20 percent of officer positions, it was a "watershed" in army history and began the process of usurpation of the militia as the central piece of the peacetime establishment. Skelton, *An American Profession of Arms*, 128–29.

33. Barnard, *Military Schools*, 732–33.

34. Couper, *Claudius Crozet, Soldier-Scholar-Educator-Engineer*.

35. For a discussion of the role of personal honor in antebellum society and the differences between Northern and Southern ideas of honor, see Wyatt-Brown, *Southern Honor: Ethics and Behavior in the Old South*, and Bruce, *Violence and Culture in the Antebellum South*. For discussion of the clashes between cadets and Thayer, see Pappas, *To the Point*, 121–28, 139.

36. Hogen, "Modes of Discipline."

37. Simon Bernard and W. M. McRee, "Consideration on the Course of Instruction necessary for the Officers of the different arms of an Army," attached to letter from John C. Calhoun, "Relating to the Military Academies," to the Hon. R. M. Johnson, chairman of the Committee on Military Affairs, 15 January 1819, in *NASP MA*, vol. 16, no. 147, 275–81.

38. *Annals of Congress*, 14th Cong., 1st sess., 3 January 1816, 449.

39. Strong, *Cadet Life at West Point*, 104; Engineer Department, *Rules and Regulations for the Government of the Military Academy at West Point*, 23 February 1820, 13.

40. See "Act of April 29th, 1812," in Peters, ed. *Public Statutes at Large of the United States of America*, 2:720–21; also Falk, "Soldier Technologist," 23.

41. McDonald, *Thomas Jefferson's Military Academy*, 119.

42. For social origins of cadets, see Skelton, *An American Profession of Arms*, 154–66.

43. For aristocracy and American colleges, see Curti, *The Growth of American Thought*, 216–17.

44. Morrison, *"The Best School,"* 101, appendixes 5 and 9.

45. Thwing, *A History of Higher Education in America*, 222–23.

46. *Superintendents' Curriculum Study for 1802–1945*, 3–5, USMA Spec. Coll.

47. Bristed, *America and Her Resources*, 344. John Bristed delivers a scathing indictment of most things American and illustrates the different trajectories that England and America were following with regard to the Enlightenment and Romanticism. See also *Superintendents' Curriculum Study*, 3–5, for limits of mathematical instruction in American civilian colleges.

48. For a comparative study of this school and early West Point, see Molloy, "Technical Education and the Young Republic."

49. Journal of Samuel Peter Heintzelman, 18 February 1825, USMA Spec. Coll.; Dennis Mahan to his mother, West Point, 24 June 1824, Mahan Papers, USMA, Spec. Coll.

50. Pappas, *To the Point*, 102–3.

51. *Annual Reports of the Secretary of War*, 1841, quoted in *Superintendents' Curriculum Study*, 9. See also *NASP MA*, vol. 16, no. 155, 307.

52. West Point remained the sole military academy until the establishment of state and private military schools in the 1840s, most of which imitated West Point directly.

53. Hart, "The American University: Its Distinctive Feature, the 'Elective System.'"

54. Barnard, *Military Schools*, 739–40.

55. Ticknor, *Life, Letters, and Journal of George Ticknor*, 1:98.

56. Ticknor, *Life, Letters, and Journal of George Ticknor*, 1:98.

57. Barnard, *Military Schools*, 739–40.

58. Calhoun, "Reduction of the Army," *ASP MA*, vol. 2, 188–89.

59. James Monroe, *Annual Message*, 1822, *NASP MA*, vol . 16, no. 154, 306.

60. Ekirch, *The Civilian and the Military*, 69; also see *Annals of Congress*, 16th Cong., 1st sess., 1627, 1630–32.

61. Huntington, *The Soldier and the State*, 28–30. For original English attitudes toward genius, see Lloyd, *History of the Late War in Germany*. For a description of variance between English and American (and continental European) methods of engineering instruction, see Shallot, "Structures in the Stream: A History of Water, Science, and the Civil Activities of the U.S. Army Corps of Engineers, 1700–1861," 16–39.

62. McDonald, *Thomas Jefferson's Military Academy*, 27.

63. Lynn, *Battle*, 192–200; Gat, *A History of Military Thought*, 269–84.

64. First Charter of the Royal Artillery Depot Woolwich, http://www.royalengineers.ca/rma.html accessed on 11 September 2009.

65. From commentary by C. Malorti de Martemont in his introduction to the first English translation of von Bülow, *The Spirit of the Modern System of War*, ii–iv.

66. Shallot, "Structures in the Stream," 38.

67. John E. Wool to J. C. Calhoun, undated, refers to an inspection report of 12 December 1819, in John E. Wool Papers, New York State Library, Albany.

68. *ASP MA*, vol. 1, no. 173, 15 January 1819, "Additional Military Academy," 834–37, and again in "Report of the Annual Board of Visitors to the U.S. Military Academy: Made to the Congress and the Secretary of War, 1821," 3, USMA Spec. Coll.

69. *ASP MA*, vol. 1, no. 173, 835–36.

70. Chief Engineer Joseph Totten to Secretary of War Jas. M. Porter, Washington, 15 July 1843, concerning the role of West Point in "the stability of the foundation of our Military Structure," written in response to Board of Visitors Report of June 1843; found in Manuscripts of the U.S. Military Academy, 1837–44, USMA Spec. Coll.

71. *Rules and Regulations for the Government of the Military Academy at West Point*, appendix D, 26.

72. Hsieh, *West Pointers and the Civil War*, 24. See also Ekirch, *The Idea of Progress in America*, 11; Johnson, *The Early Republic 1789–1829*, 69; and Howe, *What Hath God Wrought*, 630.

73. Skelton, *An American Profession of Arms*, 181–82.

74. Moten, *The Delafield Commission*, 45.

3. West Point's Scientific Curriculum

1. Barnard, *Military Schools*, 740.

2. Skelton, *An American Profession of Arms*, 173.

3. *Catalogue of Books in the Library of the Military Academy*, 1822, in the USMA Spec. Coll. See also *Regulations for the United States Military Academy*, 1824, article 186: "No cadet shall read, or keep in his room, without permission any novel, romance, or play," in *ASP MA*, vol. 2, no. 256, 657.

4. Pappas, *To the Point*, 145–46. For instance, see Records of USMA Dialectic Society, 1840–1844, USMA Spec. Coll., where is found a fascinating record of debates by future Civil War generals Longstreet, Stewart, McLaws, Ripley, Garnett, Hardee, and Van Doren on such questions as "Have the Literary Criticisms of modern periodicals exerted a beneficial influence on national literature" (17 October 1840), "Are fictitious works production of more evil than good" (12 December 40), "Whether a state has a right to secede from the union" (6 and 28 March 1841).

5. See Barnard, *Military Schools* for an outline of early nineteenth-century curricula of European schools.

6. Quoted in Shallot, "Structures in the Stream," 37.

7. *Report of the Board of Visitors*, 1821; cited as addendum to the board report in Griess, "Dennis Hart Mahan," 69; *Annual Report of the Board of Visitors to the United States Military Academy*, 1823, 3.

8. Cowley, "European Influences upon American Higher Education."

9. *Catalogue of Books in the Library of the Military Academy* (Newburgh NY, 1822), USMA Spec. Coll. The compilation of French to English texts was done by Griess, "Mahan," 71. *The Centennial*, 1:317. The academy continued with heavy emphasis on French, but as more and more texts were translated into English, the number of hours of preparation and instruction decreased to 544 and the hours of recitation to 272 by 1861, adding Spanish and riding instruction in its place.

10. Le Sage, *Histoire de Gil Blas de Santillane*, introduced into the curriculum in 1820. See Report Superintendent USMA, 1896, 138. Other texts included C. Berard,

Lecon francaises a l'usage des commencants. For identification of texts, see *The Centennial*, 1:322–23.

11. Le Sage, *Histoire de Gil Blas de Santillane.* One vignette, for instance, entitled "Courage de la femme D'un Canonnier" (page 11), briefly told the story of Molly Pitcher, the famous American military heroine of the Revolution. Another gave advice in "De L'utilite de l'histoire" (78–79); another a moral lesson from a dialogue between Alexander the Great and "Le Philosophe" in "Alexandre et le Solitaire du Mont Caucase" (84–88).Yet another outlined correct etiquette "sur La conversation" (213); and the final piece educated the cadets about the Pole Lovzinski who fought for the United States in the Revolution in "Lodoiska" (247–97).

12. Cadet Engles to Mother, 24 January 1822, Engles Papers, USMA Spec. Coll. For texts, see Hutton, *A Course of Mathematics in Two Volumes*; M. Lacroix, *Élémens de l'algèbre*; and Biot, *Élémens de la géometrié.*

13. Paine, "Of the Comparative Powers and Expense of Ships of War, Gun-boats, and Fortifications."

14. See cadet notes in H. W. Griswold Papers, F. U410.D6 G889t, USMA Spec. Coll.

15. These texts were replaced with American works in the 1830s and 1840s; see *The Centennial*, 1:439–66.

16. Voltaire, *Histoire de Charles XII, Roi de Suede*, vi–ix.

17. Voltaire, *Histoire de Charles XII*, 43.

18. Voltaire, *Histoire de Charles XII*, 176.

19. Voltaire, *Histoire de Charles XII*, 345–48.

20. The 1820s standard texts included Dr. Olinthus Gregory's *Mechanics and Practical Astronomy* and Newton's *Principia*; see *The Centennial*, 1:261–74, 439–66,

21. In the 1820s the academy used William Henry's *Elements of Experimental Chemistry and Drawing.* Thereafter multiple texts were used; see *The Centennial*, 1:348–65, 454; ASP MA, vol. 2, "Conditions of the Military Academy 1824," 660.

22. Gay de Vernon, *Treatise on the Science of War and Fortifications*, 107–8.

23. For indications of the importance of foundries, armories, and other defense infrastructure to the emerging strategy of fortified frontiers and peacetime preparations for war, see ASP MA, vol. 2, no. 180, 42; no. 212, 337; no. 235, 450; no. 236, 472; no. 238, 523; no. 242, 529; no. 243, 530; no. 248, 599; no. 263, 729; and no. 268, 834.

24. Lallemand, *A Treatise on Artillery*, 1:iv; ASP MA, vol. 1, no. 173, 834–6. See also ASP MA, vol. 2, no. 269, 7 December 1824, 699.

25. Lallemand, *A Treatise on Artillery*, 1:v.

26. Lallemand, *A Treatise on Artillery*, 1:11, 17–18, 71–72.

27. Lallemand, *A Treatise on Artillery*, 1:54–62.

28. Lallemand, *A Treatise on Artillery*, 1:298–316, 234.

29. Lallemand, *A Treatise on Artillery*, 1:361.

30. Lallemand, *A Treatise on Artillery*, 1:361–62.

31. Lallemand, *A Treatise on Artillery*, 1:64–69.

32. Lallemand, *A Treatise on Artillery*, 1:335.

33. Lallemand, *A Treatise on Artillery*, 1:336.
34. Lallemand, *A Treatise on Artillery*, 1:336–37.
35. Lallemand, *A Treatise on Artillery*, 2:7–8.
36. Lallemand, *A Treatise on Artillery*, 2:8–32. Reconnoitering included knowledge of military characteristics related to canals, arable land, bridges, brooks, woods, camps, castles and citadels, climate, coasts, defiles, fords, forts, hills, plains, mountains, roads, villages, towns, and winter quarters.
37. Lallemand, *A Treatise on Artillery*, 2:8–9.
38. All quotations in this paragraph from Lallemand, *A Treatise on Artillery*, 2: 9–10.
39. All quotations in this paragraph from Lallemand, *A Treatise on Artillery*, 2:10–11
40. *The Centennial*, 1:277.
41. Gay de Vernon's *Treatise on the Science of War* was reproduced in 1817 in New York for use at the military academy. Cadets had to purchase this book through an advance in their pay; they could subsequently resell the used copy. When the treatise was discontinued in the 1830s, the library maintained over two thousand copies of the text. Vol. 1: v.
42. Gay de Vernon, *Treatise*, 1:vii–ix.
43. Gay de Vernon, *Treatise*, 1:v.
44. Gay de Vernon, *Treatise*, 1:9–10. Here we see the emerging idea that skilled scientific officers could form an excellent general staff corps, something that would not be formally articulated by American officers until the 1840s.
45. *ASP MA*, vol. 2, no. 206, Fortifications, 308.
46. *ASP MA*, vol.1, 239
47. *ASP MA*, vol. 1, 241.
48. *ASP MA*, vol. 1, 248.
49. *ASP MA*, vol. 1, 250–58.
50. For primary sources see de Vauban, *Traité de l'attaque et de la défense des places fortes*, reproduced in part in Gérard Chaliand, *Anthologie Mondiale de la Stratégie* (Paris: Robert Lafont, 1990), 658; also see Chevalier de Cambray, *Manière de fortifier de Mr. de Vauban* (Amsterdam: Pierre Mortier, 1689) reproduced in Langins, *Conserving the Enlightenment*, 58–62. For secondary sources see Guerlac, "Vauban: The Impact of Science on War," 69; see also Duffy, *Fire and Stone*, 45–47.
51. See excellent example of student work in "Memoir on the Plan of the Front of Fortification in Horizontal Ground and Dry Ditches," Jacob Whitman Bailey Papers, USMA Spec. Coll.
52. Langins, *Conserving the Enlightenment*, 78–79.
53. Lloyd, *Vauban, Montalembert, Carnot*, 78.
54. Vauban, *Le Triomphe de la Méthode*.
55. Vauban, *A Manual of Siegecraft and Fortification*, 21.
56. Clausewitz, *On War*, 153.
57. Quoted in Chandler, *The Campaigns of Napoleon*, 135.
58. Gay de Vernon, *Treatise*, 1:106.

59. Gay de Vernon, *Treatise*, 1:106.
60. O'Connor's summary attached to Gay de Vernon, *Treatise*, 2:288, see footnote, also 64.
61. Gay de Vernon, *Treatise*, 2:387–88.
62. Gay de Vernon, *Treatise*, 2:385–416.
63. Gay de Vernon, *Treatise*, 2:386.
64. Gay de Vernon, *Treatise*, 2:431–32
65. Gay de Vernon, *Treatise*, 2:401, 413.
66. Gay de Vernon, *Treatise*, 1:102.
67. Gay de Vernon, *Treatise*, 1:104.
68. Gay de Vernon, *Treatise*, 2: 480.
69. Gay de Vernon, *Treatise*, 2:467.
70. Gay de Vernon, *Treatise*, 2: 476.
71. Skelton, *An American Profession of Arms*, 222. See also the *Army Register* for a complete listing of all army officers and their employment.

4. Internal Improvements

1. Nelson, "Military Roads for War and Peace," 1, 3.
2. Annual Message of the President, 2 December 1806, in Ford, ed., *The Writings of Thomas Jefferson*, 7:494.
3. *ASP Misc.*, vol. 1, 456–57, 724–921.
4. Hill, *Roads, Rails & Waterways*, 38.
5. Nelson, "Military Roads for War and Peace," 4–5.
6. *NASP MA*, vol. 1, Letter from John Calhoun, 20 January 1818, 83.
7. *ASP MA*, vol. 1, no. 206, "Fortifications," 15 February 1821, 305.
8. Report on a Proposed Canal through the Mining District of New Jersey, Brig. Gen. S. Bernard and Col. J. Totten, 5 November 1823, in Engineer Department, file of Reports of the Board of Internal Improvements, vol. 2, 1–27.
9. Schubert, *Vanguard of Expansion*, 1–2, chapter 1.
10. The scope of this work was anticipated as early as 1816. See Memoir of John Anderson and Isaac Roberdeau, 15 January 1816, Engineer Department, Bulky File, no. 207, RG 77, NA. Much of the data collected on these expeditions furnished material to ethnologist Henry Schoolcraft for his seminal work on "Amerindians."
11. Order to the Bureau of Topographical engineers, 3 March 1813, in Burr, "Historical Sketch of the Corps of Engineers, U.S. Army," 34. See also Goetzman, *Army Exploration in the American West*, 6–21.
12. *ASP MA*, vol. 3, no. 339, 492, submitted in 1827 by Topographical Bureau.
13. *ASP MA*, vol. 3, no. 339, 492.
14. Hill, *Roads, Railways & Waterways*, 25; Shallot "Structures in the Stream," 62
15. Engineer Department, Report on the Topographical Engineers, 25 December 1822, Bulky File, no. 114, 11–22, NA.
16. See *Acts of the Eighteenth Congress*, Doc. 35, 30 April 1824.

17. Annual Message to Congress, 7 December 1824, 18th Cong., 2nd sess., S. Doc 1, 11. Also see *ASP MA*, vol. 3, no. 281, 102.

18. Hill, *Roads, Railways & Waterways*, 60, 65

19. Letter sent by Chief Engineer Alexander Macomb, Engineer Department, Washington, 24 March 1828, Misc. Letters Sent, vol. 5, 1–3, NA.

20. *ASP MA*, vol. 3, no. 292, 184–86.

21. Hill, *Roads, Railways & Waterways*, 34–35; Endler, *Other Leaders, Other Heroes*, 29–35.

22. Couper, *Claudius Crozet, Soldier-Scholar-Educator-Engineer*, 34.

23. *Army and Navy Chronicle*, 29 January 1835; Pappas, *To the Point*, 186; Morrison, *"The Best School,"* 20.

24. *ASP MA*, vol. 3, no. 327, 278.

25. Griess, "Mahan," 171.

26. Molloy, "Technical Education in the Young Republic," 457.

27. Huntington, *The Soldier and the State*, 195–211.

28. *ASP MA*, vol. 3, no. 339, 492.

29. *ASP MA*, vol. 3, no. 327, 279. This document gives details regarding each officer on each project, esp. on 282.

30. *ASP MA*, vol. 3, no. 327, 199.

31. See Hope, "A Scientific Way of War: Antebellum Military Science, West Point, and the Origins of American Military Thought," appendixes D and E. Accessed at http://qspace.library.queensu.ca/handle/1974/7326.

32. *ASP MA*, vol. 5, no. 613, "Report from the Topographical Bureau," 2 November 1835, 715.

33. Hill, *Roads, Railways & Waterways*, 52–53.

34. Hope, "A Scientific Way of War," appendixes D and E.

35. *General Regulations for the Army of the United States, 1841*, article 921.

36. *ASP MA*, vol. 4, no. 390, 1–21, "1828 Annual Report on the Army—Its Purpose and Work." See also *ASP MA*, vol. 3, 284–85, "Revised Report of the Board of Engineers," 12 April 1826.

37. It is difficult to generalize regarding the social respectability of the army officer corps for the entirety of the antebellum period in every corner of the United States. Skelton in *An American Profession of Arms* (297–304) concludes that generally along the frontiers, where officers had to perform constabulary duties and where they often favored the claims of Indians over settlers, their popularity was low. However, in the east and especially in the cities and large towns, officers were perceived as gentlemen, trustworthy and respectable. John Schofield claimed in his 1898 biography that "before the Civil War an officer of the army needed no indorser anywhere in the country. His check or his pay account was as good as gold." Schofield, *Forty-Six Years in the Army*, 17. Edward Coffman paints a balanced impression of the officer corps in *The Old Army*, stating that respectability came when social interaction allowed civilians to see the result of an officer's work and when garrisons could enter-

tain civilians (see chapter 2). The feeling of isolation that officers sometimes expressed was largely felt in remote frontier garrisons, where they often were surrounded by settlers who resented the army as a representative of federal authority. To generalize the frontier relationship is wrong.

38. *NASP MA*, vol. 1, 187; report from Clay County Missouri, 24 December 1835, referred to the Committee of Military Affairs; Presented to the Senate and House Representatives of the United States in Congress; (quotation from page 188).

39. *ASP MA*, vol. 7, 778. See also Nelson, "Military Roads for War and Peace," 12–13.

40. *NASP MA*, vol. 1, 258–59, Letter from the Secretary of War, referred to the Committee on Military Affairs, 3 January 1838 (letter written 30 December 1837); troop estimates are on 269, estimates of Native warriors are found on 273.

41. *General Regulations for the Army of the United States*, 1841, 138–41.

42. Risch, *Quartermaster Support of the Army*, 211, 487.

43. *Annual Report of the Board of Visitors to the United States Military Academy*, 1821, 1.

44. *Annual Report of the Board of Visitors to the United States Military Academy*, 1830. See also Hill, *Roads, Railways, & Waterways*, 26.

45. *ASP MA*, vol. 3, no. 284, Annual Report of the Secretary of War 1825, 109.

46. *The Centennial*, 1:294.

47. Endler, *Other Leaders, Other Heroes*, 51; Cullum, *Biographical Register*, 1:321–22. Mahan's efforts were quickly recognized by civilian academia and in 1837 he was conferred with honorary degrees from Brown, Princeton, and later from William and Mary and Dartmouth.

48. USMA Staff Records, vol. 2 (1835–42), 195–99.

49. Field, *Forts of the American Frontier, 1820–91: Central and Northern Plains*, 25–31. For an intriguing look into the influence of Mahan and West Point on western construction of forts and domestic quarters, including the liberal use of Greek Revival and other popular design in post construction, see Kindred, "The Army Officer Corps and the Arts: Artistic Patronage and Practice in America, 1820–85," 130–77.

50. Morrison, "Military Education and Strategic Thought," 120, 125; and Moten, *The Delafield Commission*, 365–36. For an accurate assessment of Mahan's methodology, see Griess, "Mahan," 171–208, 249–86.

51. Griess, "Mahan." Also see Chief Engineer Totten to Secretary of War Poinsett, Washington, 27 February 1841, in Manuscripts of the U.S. Military Academy, 1837–44, USMA Spec. Coll. In this letter, Totten argues that they cannot introduce more practical instruction at West Point without displacing necessary academic instruction. Totten believed that post, garrison, and field appointments would give the graduates sufficient opportunity to address deficiencies in practical exercises, but after leaving West Point there would be no further opportunity for theoretical instruction.

52. For criticism of Mahan's course as pure theory, see Morrison, *"The Best School,"* 97; also see Shiman, "Engineering Sherman's March," 38. For an overwhelming rebuttal, see Griess, "Mahan," 171–209.

53. Mahan's nasal infections gave his rendering of *common sense* the sound of "cobben sense," and thus cadets nicknamed him "Old Cobben Sense." See Shiman, "Engineering Sherman's March," 39.

54. Shallot, "Structures in the Stream," 16–19, 22.

55. Dennis Hart Mahan to editor of the *Army and Navy Journal* 4 (3 March 1866): 422.

56. Shiman, "Engineering Sherman's March," 40; Griess, "Mahan," 140–43, 181–84, 217–18.

57. Skelton, *An American Profession of Arms*, 167–72.

58. Morrison, *"The Best School,"* 15. Morrison dismisses any idea of strategic thinking in the antebellum period or the army's seriousness about its role in peacetime preparations for war. For explanation of reasoning behind the War Department policy of allocating commissions in corps of engineers to the top graduates of West Point, see Secretary of War Poinsett to the Honorable John McKeon, 20 January 1837, in Office of the Secretary of War, Letters Sent, Military Affairs, 17:95 RG 107, NA.

59. Hope, "A Scientific Way of War," appendixes D and E.

5. Jacksonian Military Science

1. Alex de Tocqueville made the observation clearly in a chapter dedicated to explaining why democratic armies prefer war; see Tocqueville, *Democracy in America*, 1:228, 2:270, chapters 22–26. See also Ekirch, *The Civilian and the Military*, 72–74.

2. White, *The Jacksonians*, 187–88; Huntington, *Soldier and the State*, 224; Utley, *Frontiersmen in Blue*, 158.

3. Only on one instance was particular funding denied. In 1844 Congress withheld funds to send the annual Board of Visitors to the academy. In lieu of this, Gen. Winfield Scott went instead with a group of officers to conduct his own board.

4. Pappas, *To the Point*, 185.

5. Morrison, *"The Best School,"* 26–27. A relationship, or at least an acquaintanceship, is implied in the letter from Jackson to Mahan, 25 October 1844, in which Jackson states: "I with pleasure, feeble as I am, send you my autograph and tender to you my best wishes—may you live a long and useful life and then enjoy a happy immortality, Very Respectfully Yours, Andrew Jackson." USMA Spec. Coll.

6. For a good rendering of Jackson's attitude toward military affairs, see White, *The Jacksonians*, 187–212.

7. *North American Review* 61 (1845): 321; quoted in White, *Jacksonians*, 191.

8. Adjutant General's Office, *Army Register* for 1840, 1845, and 1850.

9. Report from the Secretary of War, 3 December 1836, *Army and Navy Chronicle* 3, no. 25 (22 December 1836): 385.

10. Estimates are from "Foreign Intelligence," *Army and Navy Chronicle* 5, no. 11 (14 September 1837): 173.

11. The influence of the War of 1812 and the basis of the defense planning scenario were published in 1838 for public consumption: "Admonished by the diffi-

culties ... and the incredible expenses incurred at the commencement of the last war [1812], from want of preparations, the nation, immediately ... adopted measures of defence of the seaboard. The circumstances attending the contest had shown that the whole extent of the coast was vulnerable, and that an active and enterprising enemy, with comparatively small means, might keep the country in a state of alarm, by threatening the entire line, and selecting the unguarded points to attack—a mode of warfare which obliged the Government to incur great and disproportionate expense, and to employ an immense force in ineffectual attempts to protect its defenceless and extensive maritime frontier." An answer was to build fortifications capable of "effectual resistance for five or six days, against an attacking force of 10,000 men, furnished with all the means and appliances to conduct a siege.... From these dangers, neither ships of war nor steam batteries can protect the nation." See "Report from the Secretary of War in Compliance with a Resolution of the Senate of the 14th October, 1837," delivered 10 January 1838, NASP MA, vol. 1, 275–78.

12. Original Third System thinking regarding the Atlantic threat dominated this era. See Report by Mr. R. M. Johnson, from the Committee of Military Affairs, entitled "Defence of Patapsco River and Baltimore," NASP MA, vol. 1, 4 March 1834.

13. *Army and Navy Chronicle* 5, no. 15 (12 October 1837): 238.

14. Prucha, *The Great Father*.

15. Report of the Secretary of War Indian Affairs, *Army and Navy Chronicle* 3, no. 26 (29 December 1836): 401–2.

16. Annual Report War Department, 1828; Annual Report on the Army, ASP MA, vol. 4, no. 390, 1–2; also see Annual Report of the Secretary of War (tabled 2 December 1837), in *Army and Navy Chronicle* 5, no. 24 (14 December 1837): 369. Indian removals soured many army officers; see Howe, *What Hath God Wrought*, 750; also Winders, *Mr. Polk's Army*, 34.

17. For a list of projects, see Morrison, *"The Best School,"* 9; and *The Centennial*, 1:486–87.

18. Report of the Secretary of War, *Army and Navy Chronicle* 3, no. 25 (22 December 1836). In 1836, 24,500 non-regular soldiers were engaged under both federal and state sponsorship to fight Indians and to guard frontiers. To rely on militia alone was considered foolhardy, as one commentator in *Army and Navy Chronicle* opined about militia training days: "Nobody can have forgotten the scenes of riot, brawling, and debauchery, which were almost universally exhibited at public reviews and battalion trainings. The dregs of society were always congregated in great numbers to get drunk and fight.... These trainings, in fact, were a public nuisance, a disgrace to the State; their demoralizing consequences cannot be estimated." Found in "The Militia Law," *Army and Navy Chronicle* 1, no. 19 (7 May 1835): 1. Again, in the 17 September edition, under the title "Militia System," the militia system was criticized as "worse than useless." *Army and Navy Chronicle* 1, no. 38 2 (17 September 1835): 299.

19. Ekirch, *The Civilian and the Military*, 79.

20. Report of the Secretary of War of 7 April 1836, *Army and Navy Chronicle* 2, no. 26 (30 June 1836): 401–2.

21. *NASP MA*, vol. 1, 258–59, Letter from the Secretary of War, referred to the Committee on Military Affairs, 3 January 1838 (letter written 30 December 1837); troop estimates are on p. 269, estimates of Native warriors are on p. 273.

22. From a letter from Brigadier Gaines to Secretary of War Poinsett, 19 August 1838; quoted in Johnson, *Winfield Scott*, 20.

23. *NASP MA*, vol. 2, 91, 97–98, Memorial of Edmund P. Gaines, 6 March 1840.

24. Gaines to Poinsett, 6 May 1839, Adjutant General's Office, Letters Received 1839, RG 94, NA.

25. *NASP MA*, vol. 1, 288, Letter from Brigadier Gaines to the Honorable L. F. Hinn and A. G. Harrison, St. Louis.

26. *ASP MA*, vol. 4, no. 407, 106. This was not just a matter of liking cavalry but a practical issue of most infantry and artillery being located 300 to 1,200 miles from key seacoast locations, too far away from possible attack "by the *regular troops* of a civilized nation."

27. This act prohibited officers from being separated from their line companies or regiments to conduct internal improvements work, as these separations, often for years, were eroding cohesion in the line units. See White, *The Jacksonians*, 199.

28. Traas, *From the Golden Gate to Mexico*, 66.

29. One appendix was 25 pages in length on botany of the Southwest. Another had 16 pages of meteorological observations. Another was dedicated to geographic distances. One appendix contained 205 pages of astronomical data containing more than 2,000 astronomical observations of precise locations and 357 barometric observations that showed elevations en route. *Emory, Notes of a Military Reconnaissance*, reproduced in part in Traas, *From the Golden Gate to Mexico*, 83–87.

30. *NASP MA*, vol. 1, 275–80, Report from the secretary of war in compliance with a resolution of the Senate of the 14th October 1837, delivered 10 January 1838.

31. Board of Engineer Report Defense of the Northern Frontier, 27 December 1837, Joseph Swift Papers, USMA Spec. Coll.

32. Lt. Miner Knowlton, Report on the Military Establishments of the British Provinces 1840, West Point, December 1840, USMA Spec. Coll.; Joseph Totten to Richard Delafield, Washington, 31 March 1843, Records of the Office of Chief of Engineers, entry 146, RG 77, NA.

33. Secretary of War Porter to President Jackson, 25 February 1829, in *ASP MA*, vol. 4, no. 407, 106.

34. Memorial of Edmund P. Gaines, 6 March 1840, in *NASP MA*, vol. 2, 91: also see Skelton, *An American Profession of Arms*, 244, and Watson, "Knowledge, Interest, and the Limits of Military Professionalism," 293. Gaines was supported by Alden Partridge. See Memorial of a Committee of the Military Convention at Norwich Vermont, 24 February 1840, in *NASP MA*, vol. 2, 69. Partridge was against the "useless system of fortifications" and challenged the "charm in the word *forti-*

fication, which pervades and powerfully influences both the people and their representative," and the word *permanent*, which he felt made the system "irresistible." He wanted to reduce the system to just a couple of permanent fortifications. Two-part report found in *Army and Navy Chronicle* 2, no. 18 (5 May 1836): 273–77, no. 19 (12 May 1836): 289–92. Gaines wanted steam batteries instead of permanent fortifications wherever they suited, such as on the Delta of the Mississippi. He gave details of exact numbers of guns required to achieve a specific effect of fire at specific locations, such as at Hampton to protect the Hampton Roads and Norfolk. Gaines challenged the board of engineers' planning assumptions regarding an invasion of 20,000 enemy soldiers, calling it too hypothetical. He wanted more planning using means other than permanent works. For the defense of Washington, for example, he did not support more fortifications on the Patuxent River: "Our navy, our floating batteries, our means of communications and concentration, seem to me far better adapted to the defence of this city, than forts at the distance of nearly fifty miles" (291).

35. *Army and Navy Chronicle* 2, no. 19 (12 May 1836): 291–92; see also NASP MA, vol. 1, 240–41.

36. Memorial of Edmund P. Gaines, 6 March 1840, NASP MA, vol. 2, 91.

37. The details of the Third System were made available to the public, published in records of House debates and in journals such as *Niles' Weekly Register* and the *Army and Navy Chronicle*. Complete reports detailing location of works, expenditures, armaments, and garrisons were published every year. There is evidence that these reports did indeed shape public thinking on defense. In the scare of war with England in 1841, for example, a memorial from the corporate authorities of the city of Baltimore lobbied Congress for 159 guns and a 800-man garrison at Soller's Flats, a stated requirement of the Third System. Report submitted to the Senate of the United States, 2 March 1841, in NASP MA, vol. 2, 155.

38. *Annual Report of the Board of Visitors to the United States Military Academy* 1830, in ASP MA, vol. 4, no. 458, 607.

39. The 1832 curriculum remained unchanged until 1854. See Griess, "Mahan," 177 and appendix B.

40. Academic Board records reveal the frequency, depth, and details of the problem of adding things to the curriculum. See USMA Staff Records, vol. 2, 15 September 1838, 112–16; vol. 3, 18 October 1843, 194–209; vol. 4, 20 January 1845, 15–28, 10 February 1845, 36–52; vol. 5, 28 July 1854, 440–44; vol. 6, 4 December 1856, 188.

41. *Annual Report of the Board of Visitors* 1830, 607.

42. *Annual Report of the Board of Visitors* 1830, 607.

43. *Army and Navy Chronicle* 5, no. 1 (6 July 1837): 2.

44. Skelton, *An American Profession of Arms*, 168–69.

45. Moten, *The Delafield Commission*, 55. This is an example of Isaiah Berlin's opinion that an affinity for romanticism may prevail over our appreciation of the past. It reflects postmodernist perspectives about science, a quiet discomfort regarding

the notion of immutable principles, and an admonishment that antebellum instructors avoided dialectical reasoning.

46. Chief Engineer Joseph G. Totten to Hon. Jas. A. Black, Washington., 26 February 1844, 5, in Manuscripts of the U.S. Military Academy, 1837–44, USMA Spec. Coll., 10–11. See also Cajori, *The Teaching and History of Mathematics*, 84–86, 114–27. It must be added that colleges in the United States, after initially following British practices of rejecting mathematics in undergraduate curriculae, steadily increased mathematical instruction throughout the nineteenth century.

47. Totten to Secretary of War Jas. M. Porter, Washington, 15 July 1843, concerning the role of West Point in "the stability of the foundation of our Military Structure," written in response to Board of Visitors' Report of June 1843; found in Manuscripts of the U.S. Military Academy, 1837–44, USMA Spec. Coll.

48. *Army and Navy Chronicle* 5, no. 1 (6 July 1837): 1–6.

49. Totten to Black, 26 February 1844, 5–6.

50. Totten to Black, 26 February 1844, 6.

51. Totten to Black, 26 February 1844, 7.

52. Totten to Black, 26 February 1844, 11.

53. Adjutant General's Office, *Army Register*, 1844.

54. Totten to Black, 26 February 1844, 12. Totten also stated his belief in West Point as a national institution and that its patriotism and singular education remained with the graduate for life. He insisted that regular officers learned to be "beyond the reach of all party influences or sectional prejudices" (18).

55. Totten to Black, 26 February 1844, 23.

56. Skelton, *An American Profession of Arms*, 169.

57. Maj. Sylvanus Thayer to Maj. Richard Delafield, 1 October 1838, Richard Delafield Papers, USMA Spec. Coll.

58. Totten to Delafield, 18 November 1842, in Adjutant General War Department (AGWD), USMA, 1812–67, vol. 10, 300–305, USMA Spec. Coll.

59. Totten to Poinsett, Washington, 27 February 1841, in Manuscripts of the U.S. Military Academy, 1837–44, USMA Spec. Coll.

60. Annual Report of the Board of Visitors to the U.S. Military Academy, 1836, *Army and Navy Chronicle* 3, no. 1 (7 July 1836): 3–4.

61. Totten to Mahan, Washington, 26 January 1839, USMA Spec. Coll.; Totten to Porter, 15 July 1843, concerning "The Stability of the Foundation of Our Military Structure" in response to the "very emphatic testimony given by the Board as to the necessity of incorporating with the course of practical instruction, the exercise of sapping, Mining and Military Bridging, will produce the merited effect." This was a recurrence. The 1841 Board of Visitors recommended the same to Secretary of War John Spencer; see also Totten to Poinsett, Washington, 27 January 1842, Office of Chief Engineer, Letters Sent, Abstract of Letters Sent, 1 and 10:204; also Report of the Committee of the Academic Board, February 1842, USMA Staff Records, vol. 2 (1835–42).

62. *The Centennial*, 1:415–16.

63. Mahan to Secretary of War, 1 August 1840, Poinsett Papers, 15:11, Historical Society of Pennsylvania.

64. *ASP MA*, vol. 4, no. 407, 106.

65. All quotations in this paragraph come from Mahan, *Treatise on Field Fortifications*, ii–viii.

66. Mahan, *Treatise on Field Fortifications*, xv. Mahan believed that a combination of fear and inability to estimate ranges had caused most shots fired to miss their mark. Field fortifications worked to reduce this fear, but officers needed to be taught the mathematics of range estimation.

67. Mahan gave rough calculations for practices, but encouraged practical applications. See Mahan, *Treatise on Field Fortifications*, xvi, xvii, xx–xxi, xxvi.

68. Mahan, *Treatise on Field Fortifications*, 2–4, 6–10.

69. Mahan, *Treatise on Field Fortifications*, 19.

70. Mahan, *Treatise on Field Fortifications*, 135, 145.

71. Mahan, *Treatise on Field Fortifications*, 73–74. Mahan cited this from Rogniat, *Considérations sur l'art de la guerre*. Baron Joseph Rogniat was a chief of the French engineer corps.

72. Gay de Vernon, *Treatise on the Science of War*, 185–94; Lallemand, *Treatise on Artillery*, 130.

73. Lallemand, *Treatise on Artillery*, 131–32. Specifically Mahan felt that a topographical engineer "should unite a large fund of general information on statistics and the natural sciences; a fund which nothing but a life of daily study and observation can supply."

74. See Mahan, "Composition of Armies," USMA Spec. Coll., 31–32.

75. Mahan to R. F. Buller, Secretary of War, West Point, 6 February 1837, USMA Spec. Coll. Mahan claimed that his work was "of a strictly practical character." He realized the need to disseminate knowledge to the militia, who would rely on field forts in a large-scale war. Mahan dedicated the book to militia officers in hopes of "strengthening the Army . . . by raising the standard of military information among the militia." See also Cullum, *Biographical Register*, 1:321; Griess, "Mahan," 303–4; Cox, *Military Reminiscences of the Civil War*, 1:30; same found in the 2008 reprint by Bibliobazaar on 1:41; see also *Army and Navy Journal* 1 (19 December 1863): 265; Mahan to editor in "West Point and Horse Sense," *Army and Navy Journal* 3 (31 March 1866).

76. USMA Staff Records, vol. 2 (15 September 1838): 112–16; vol. 3 (18 October 1843): 194–209; vol. 4 (20 January 1845): 15–28.

77. For details of Mahan's course in military engineering and the science of war, see USMA Staff Records, vol. 2 (1835–42): 192–95, USMA Spec. Coll. Also see Griess, "Mahan," 166–67, 171–81, 224.

78. Mahan, "Composition of Armies," 2.

79. This was not mere nuance. Jomini was becoming widely read in Europe when Mahan started lecturing and publishing. Had he been a mere copycat, Mahan

would have repeated Jomini verbatim, as was the practice in much of Europe. See Jomini, *An Exposition of the First Principles of Grand Military Combinations and Movements*, 7–48.

80. Mahan, *"Composition of Armies,"* 3.
81. Mahan, *"Composition of Armies,"* 3.
82. Mahan, *"Composition of Armies,"* 5.
83. Hittle, ed., *Jomini and His Summary of the Art of War*, 9–10.
84. See Morrison, *"The Best School,"* 96–97. Morrison has stated, "It seems highly unlikely that one nine-hour period out of entire four-year program would so impress a student that it would continue to govern his thinking years later." This assertion is now so widely repeated it has become a big part of the accepted interpretation that West Pointers were unprepared when they assumed high command during the Civil War. His analysis of that curriculum is incorrect and failed to see connections between the various subjects as components of a coherent doctrine all related to strategy. He attributes no relationship to Vauban's fortification theory or Vauban's "method" and campaign planning. For the most comprehensive coverage of the Mahan course, see Griess, "Mahan," 174–78, 180, 197, 223, 362–66.
85. USMA Staff Records 2 (1835–42): 112–16, USMA Spec. Coll.
86. Griess, "Mahan," 232. Also from draft letter Delafield to Totten, presented to Academic Board, 12 February 1845, USMA Staff Records, vol. 4 (1845–50), 42–43.
87. Mahan to Captain Thomas, West Point, 14 December 1844, in which Mahan states he is considering moving field fortifications to another year and stone cutting and parts of permanent fortifications as well. He wishes to begin forty-five hours on the science of war and ten on offensive and defensive operations. USMA Spec. Coll.
88. This was in part considered because of a recommendation made by the select committee on the subject of the Military Academy at West Point to form there a school of application for serving Army officers; see proposal of 1 March 1837 by Mr. Smith under title "Military Academy," in *Army and Navy Chronicle* 4, no. 9 (2 March 1837): 141.
89. *Annual Report of the Board of Visitors*, 1843. 1.
90. Totten to Secretary of War James Porter, 15 July 1843, AGWD, on USMA, 1812–67, vol. 27; Totten to Delafield, 27 July 1843, USMA Staff Records, vol. 3 (1842–45), 166, USMA Spec. Coll.
91. USMA Staff Records, vol. 3 (1842–45), 194–209, USMA Spec. Coll.
92. Letter, Totten to Delafield, 6 December, 1844, AGWD, on USMA, 1812–67, vol. 12, 247.
93. Report from the Mahan Committee to the Academic Board, 20 January 1845, USMA Staff Records, vol. 4 (1845–50), 22, USMA Spec. Coll.
94. This was not, of course, military history as we know it today. It gave more attention to narrative of campaigns and generals than to historical analysis, except when applying principles to past campaigns, which Mahan favored. For general desire to see history taught, see "Military Academy," *North American Review* 34 (January

1832): 260; *Annual Report of Board of Visitors*, 1838, 1854, U.S. Senate, 33rd Cong., 2nd sess., Doc. 1, 130, 151.

6. Military Science during and after the Mexican War

1. Millett and Maslowski, *For the Common Defense*, 144.
2. There can be no doubt about the initial campaign design using the Brazos Island as a base of operations. See Polk's message to the House on 11 May 1846 in *Messages of the President of the United States: with the correspondence, therewith communicated, between the Secretary of War and other officers of the government, on the subject of the Mexican War*, 7, and orders to Taylor, 81–83.
3. Secretary of War Marcy to Brigadier Kearney, 3 June 1846, *Messages of the President*, 153.
4. White, *The Jacksonians*, 50–67.
5. Most militia units were sent home quickly; see Williams, *Americans at War*, 37. States raised volunteer units quickly, but their quality varied, the worst being a "lawless drunken rabble." Nonetheless, the mobilization of volunteers was necessary to avoid a repeat of the War of 1812 when states' militias refused to cross the border into Canada. To be fair, this recurred in 1846, with the New England states boycotting support for the war. See Bernardo and Bacon, *American Military Policy*. The expansion of the regulars was achieved by filling up regiments that had been kept at half strength of private soldiers, thus doubling the 8,600-man army in 1846 without a decrease in effectiveness.
6. Howe, *What Hath God Wrought*, 750.
7. Wilson, *The Business of Civil War*, 44.
8. In *Messages of the President*, see messages of Brig. R. Jones, adjutant general, 6 August 1845, regarding the concentration of soldiers for General Taylor (83–84) and the ships made available for communications to the seat of war (85–86).
9. *The Centennial*, 1:601, 603, 605.
10. Traas, *From the Golden Gate to Mexico*, 118–22. See also message from Secretary of War Marcy and President Polk to Brigadier Taylor directing him to reconnoiter the frontier and gain intelligence on Mexican forces, in *Messages of the President*, 88, and directing that engineers and Topogs supply information and reconnoiter locations for defensive position along the frontier, 89–90.
11. Cullum, *Biographical Register*, 1:543–45.
12. *Messages of the President*, 25 March 1846, 129–30.
13. Grant, *Personal Memoirs*, 1:39. This claim is not supported by academy library circulation records, which show no withdrawals of novels but several of the works of Livy. Grant practiced a trait known to many general officers in retirement, remaining disparaging about their own intellectual inclination toward academic learning in general. It is still trendy for general officers visiting their alma mater to make remarks about how during the visit they finally found where the library was located. Historians must be cautious about memoirs that paint anti-intellectual self-portraits.

14. Despite Grant's later dismissal of his own academic prowess, letters from his cadet days reveal an intense interest in mathematics and academic inclination. See Grant to his cousin, Camp Biddle West Point, 18 July 1840, USMA Spec. Coll.

15. Grant, *Personal Memoirs*, 1:40, 51–52.

16. Taylor named it Fort Taylor; however, its nickname, Fort Texas, became better known. After the death of its commander, Major Brown, on 9 May 1946, it became Fort Brown; see Brigadier General Taylor to Adjutant General Jones, 6 April 1846, in *Messages of the President*, 133.

17. *The Centennial*, 1:607–8.

18. Traas, *From the Golden Gate to Mexico*, 132–33. Taylor used his West Point graduates to help formulate his plans. See Grant, *Personal Memoirs*, 1:109–10.

19. Grant, *Personal Memoirs*, 1:97.

20. See report of Captain Hughes of Wool's march into Mexico in Traas, *From the Golden Gate to Mexico*, appendix A, 229–303.

21. The bridge was invented by Capt. J. F. Lane in 1834; see *Army and Navy Chronicle* 3, no. 15 (13 October 1836): 254, and no. 17 (3 November 1836): 274.

22. G. Smith, "Company A Engineers in Mexico, 1846–1847," 47–58. Smith conducted the basic infantry skills instruction of this company, and Swift did the engineering instruction "in the rudiments of practical military engineering which he had acquired at Metz" (48). The company set sail from New York on 26 September 1846 with seventy-one soldiers.

23. Smith, "Company A Engineers; Myers, ed., *The Mexican War Diary of General George B. McClellan*.

24. The company commenced on 21 December 1846 and crossed an incredible complex of streams, wadis, and steep river valleys before joining Taylor in Victoria in January 1847. One particular stream was a hundred foot drop on the near bank, a hundred feet across the stream, and a hundred feet straight up the near bank of mud. Smith used eight hundred men with two hundred tools to build ramps down, across, and up the other side in eight hours.

25. Myers, *The Mexican War Diary*, 16–17. McClellan's account provides even greater detail regarding the engineer work in road building during this advance.

26. Traas, *From the Golden Gate to Mexico*, 142, 156–57, 160.

27. Wilcox, *History of the Mexican War*, 261. See also *The Centennial, 1:*614.

28. Williams, *With Beauregard in Mexico*, 9.

29. Wilcox, *History of the Mexican War*, 261.

30. Williams, *With Beauregard in Mexico*, 11.

31. Myers, ed., *The Mexican War Diary*, 62–64.

32. Traas, *From the Golden Gate to Mexico City*, 200.

33. Traas, *From the Golden Gate to Mexico City*, appendix C, 320.

34. *The Centennial*, 1:616; Williams, *With Beauregard in Mexico*, 36; and Myers, ed., *The Mexican War Diary*, 80–81.

35. *The Centennial*, 1:619.

36. Scott's report from Mexico City, 18 September 1847, in *The Centennial*, 1:624–26.
37. Williams, *With Beauregard in Mexico*, 46.
38. Grant, *Personal Memoirs*, 1:145–46.
39. Traas, *From the Golden Gate to Mexico*, 209; see also Steele, *American Campaigns*, 1:122.
40. Howe, *What Hath God Wrought*, 749; *The Centennial*, 1:628.
41. Abbot, "Memoir of Dennis Hart Mahan," 2:33.
42. Millett and Maslowski, *For the Common Defense*, 134–35.
43. Skelton, *An American Profession of Arms*, 221–37.
44. Skelton, *An American Profession of Arms*, 234–36. What made matters more difficult for commanding generals and some politicians was the long tenure of service of staff department chiefs. Thomas Jessup served as quartermaster of the army from 1818 until 1860. Roger Jones was adjutant general from 1825 to 1852. George Crogan was the inspector general from 1825 to 1845. Thomas Lawson was surgeon general from 1836 to 1861. Joseph Totten was chief engineer from 1838 to 1864. John Abert was chief of topographical engineers from 1834 to 1861. J. E. Lee was judge advocate general from 1849 to 1862. George Talcott was the chief ordnance from 1851 to 1861.
45. Secretary of War, Annual Report (1854), 33rd Cong., 2nd sess., Sen. Exec. Doc. 1, vol. 2 (serial 747), 11–16; Secretary of War, Annual Report (1855), 34th Cong., 1st sess., Sen. Exec. Doc. 1, vol. 2 (serial 811), 6; Secretary of War, Annual Report (1859), 36th Cong., 1st sess., Sen. Exec. Doc. 2, vol. 2 (serial 1024), 4–5; also Utley, *Frontiersmen in Blue*, 51; and Wilson, *The Business of Civil War*, 40–41. Davis's suggestion was never realized because it was politically impractical. What was needed most of all was forced retirements of senior staff officers who had been in Washington for decades. This happened during the first year of the Civil War.
46. Mahan to Maj. Gen. George McClellan, 1 August and 14 August 1861, quoted in Myers, *The Mexican War Diary*, 203.
47. Halleck, *Elements*, 239–40.
48. Skelton, *An American Profession of Arms*, 232–33.
49. Skelton, *An American Profession of Arms*, 225.
50. District and department commanders maintained a fair degree of independence, a natural thing considering distances from authority. This caused friction over lines of authority regarding attached staff officers and even troops. See Skelton, *An American Profession of Arms*, 233–34. Given this degree of independence, general staff control was even more important for the maintenance of standards during the antebellum era.
51. Moten, *The Delafield Commission*, 182.
52. The *Army Register* for 1850 informs us that 174 out of 220 designated general staff officers were graduates of the military academy. In addition to these, another 86 officers that year were borrowed from line units to fulfill general staff functions,

meaning that 306 out of 884 commissioned officers (well over 30 percent) were providing general staff for the army.

53. While the War Department took pains to ensure efficient staff capability, what suffered for this effort was the soldiery of the army. *Army Register* records make clear that during most of the antebellum period, one-quarter to one-third of the junior officers of the army were performing detached service from their commands. In 1835, for example, only 188 of 421 company officers were present with their companies. Without commanding officers, soldiers were subject to greater boredom and poorer living conditions when out on the frontier. This resulted in high rates of alcoholism and desertion. See also Utley, *Frontiersmen in Blue*, 40–41, 51–52; Skelton, *An American Profession of Arms*, 222.

54. For an examination of romanticism, see Parrington, *The Romantic Revolution in America, 1800–1860*.

55. Howe, *What Hath God Wrought*, 630–36 and chapter 16.

56. Skelton, *An American Profession of Arms*, 176.

57. Griess, "Mahan," 245, originally from *Letter to the Honorable Mr. Hawes: In Reply to His Strictures on the Graduates of the Military Academy*, by a Graduate (New York: Wiley and Long, 1836), signed "Justitia." Library copy at USMA Spec. Coll. signed by Mahan.

58. Mahan to the Napoleon Club, undated, Dennis Hart Mahan Letters, USMA Spec. Coll.

59. Griess, "Mahan," 237.

60. Careful search of the USMA's record of circulation for the 1840s and 1850s reveals which instructors were preparing such papers and evidence in the number and titles of books regarding Napoleonic campaigns lent out to an officer for a period of two to three months. See Library Circulation Records, FU510.V5 UG16, USMA Spec. Coll.

61. Griess, "Mahan," 237. See draft letter Mr. Bowman to President Lincoln, dated May 1863, USMA Letters received 1840–90, USMA Spec. Coll.

62. Maury, *Recollections of a Virginian in the Mexican, Indian, and Civil Wars*, 50–51.

63. Napoleon Club Readings, USMA Spec Coll. Speculations regarding the possible members is taken from *Officers and Members of the West Point Army Mess from 1841–1880* (in MHI) and the circulation records of the USMA Library, USMA Spec. Coll.

64. Mahan to Napoleon Club, undated, Mahan Papers, USMA Spec. Coll.

7. Antebellum Military Science

1. Skelton, *An American Profession of Arms*, 238–59.
2. Cullum, *Biographical Register*, 1:142. See Maj. W. H. Chase to Col. Joseph Totten, 17 February, 25 May, and 10 August 1846, in Records of the Office of Chief of Engineers, RG 77, NA ; William H. Chase, "Brief Memoir Explanatory of a New Trace of a Front of Fortifications in Place of the Present Bastioned Front (New Orleans, 1846)," in Report of Major W. H. Chase to Congress, 17 April 1851, Rep.

86, 37th Cong., 2nd sess., 23 April 1862. The author advocated adoption of Montalembert's system of casemated towers with tiers of cannon—over the corps of engineers' advocacy of Vauban's horizontal trace. Chase criticized Mahan's teaching. See also Griess, "Mahan," 271.

3. Sanders, *Memoirs on the Military Resources of the Valley of the Ohio.*

4. Griess, "Mahan," 270–73; Lt. James St. Clair Morton, "Memoir on the Dangers and Defences of New York, Addressed to the Hon. John B. Floyd, Secretary of War," 35th Cong., 2d sess., Sen. Exec. Doc. no. 1, 494–581.

5. Halleck, *Elements*, 37–38. Here Halleck set down the antebellum problem of definition. Admitting that war was both art and science, he surrenders to "popular language" where "it is usual to speak of the *military art* when we refer to the general subject of war, and of the *military sciences* when we wish to call attention more particularly to the scientific principles upon which the art is founded."

6. Halleck, *Elements*, 38.

7. Halleck, *Elements*, 49–56.

8. Marmont and Macdougall were two of the most influential military authors of this period, and they are remarkable here for their emphasis on field fortifications and faith in Jomini's strategic theory. See Marmont, *The Spirit of Military Institutions*, and Macdougall, *The Theory of War, Illustrated by Numerous Examples from Military History.*

9. Auguste Frederic Lendy, *The Principles of War; or, Elementary Treatise on the Higher Tactics and Strategy (1853)*, a Kessinger Publishing's Legacy Reprint (Whitefish MT: Kessinger, 2014). Mahan was influenced by Dufour's work in his 1840 lithograph notes, and he copied Dufour verbatim in his 1863 version of *Advance-Guard*, 169–305. See also Dufour, *Cours de Tactiques*, and Dufour, *Strategy and Tactics*. Mahan's plates are almost identical.

10. Halleck, *Elements*, 229.

11. Halleck, *Elements*, 67–68.

12. Halleck, *Elements*, 90.

13. Halleck, *Elements*, 97–105.

14. Halleck, *Elements*, 137, 141, 143.

15. Halleck, *Elements*, 145–48.

16. Halleck, *Elements*, 148–49.

17. Halleck, *Elements*, 151–52.

18. Halleck, *Elements*, 164.

19. He begins again by using historical examples to show how naval power in Europe, between the years 1790 and 1815, was thwarted by fortifications (164). Halleck then gives tables of costs of repairs to ships of the Royal Navy, 1800–1818, stating that Britain spent an average of $4,560,158 on timber and $4,273,371 on ship repairs each year between 1800 and 1820. Halleck then compared U.S. Navy expenditures of $5,000 to $6,600 per gun with the cost of permanent fortifications at $3,000 per gun (204–6).

20. He felt that the optimum ratio for the army should be 5 staff officers, 65 administrative services personnel, 90 artillery, and 60 engineers per 1,000, with 650 infantry and 130 cavalry. See Halleck, *Elements*, 321–23.

21. Halleck, *Elements*, 243–44.

22. Halleck borrowed this from William Duane, whom he admired. *Elements*, 246.

23. He suggested that strategy was best learned from "didactic works" and from military histories, and he listed the best readings. The theoretical works he recommended were the treatises of Archduke Charles (*Principes de la Stratégie*) and General von Wagner (*Grundsätze der Strategie*) as the best, and Jomini's *Précise de l'art de la guerre* as "exceedingly valuable." He also listed those of Rocquancourt (*Cours Elémentaire d'Art et d'histoire militaire*), Jacquinot de Presle, and Gay de Vernon. For histories he lists those of Templehoff and Lloyd (*History of the Seven Years' War*), and Montholon's *Mémoires de Napoleon*. According to Halleck, the only English history of worth was Napier's *Peninsular War*. Halleck included Clausewitz's *On War* for his chapters on strategy and the offensive—but for nothing else. It is doubtful that Clausewitz's obtuse reasoning would have appealed to Halleck. *Elements*, 58–59.

24. Totten, *Report of General J. G. Totten, Chief Engineer, on the Subject of National Defences*, 1 November 1851.

25. Totten quoted from 10 May 1840 report on national defense from board of officers to the secretary of war, H. Doc. 206, 26th Cong., 1st sess.; Totten, *Report of General J. G. Totten, Chief Engineer, on the Subject of National Defences*, 3–4.

26. Totten, *Report of General J. G. Totten*, 5–6.

27. Totten, *Report of General J. G. Totten*, 7–11.

28. Totten, *Report of General J. G. Totten*, 5, 21.

29. Totten, *Report of General J. G. Totten*, 19, 22–23.

30. Totten, *Report of General J. G. Totten*, 44.

31. Lewis, *Sea Fortifications*, 43.

32. Secretary of War, Report, 3 December 1855, Senate Exec. Document no. 78, 33rd Cong., 1st sess., vol. 7, 8–10. See also O. E. Hunt, "Federal Railways," in Miller, ed., *Photographic History of the Civil War*, 5:274.

33. Bernardo and Bacon, *American Military Policy*, 184; Adjutant General's Office, *Army Register*, 1850, 39.

34. Stunkel, "Military Scientists of the American West"; Goetzmann, *Army Exploration in the American West, 1803–1863*, 262–304.

35. *Army Registers*, 1844, 1848, 1850, 1853, and 1855. See also Utley, *Frontiersmen in Blue*, 52–58; Field, *Forts of the American Frontier, 1820–91: Central and Northern Plains*, 5; for influence of Mahan on western fortifications, see Field, 25–32. Fort design was not ad hoc, and the construction of forts, roads, and railways followed the civil and military engineering principles taught in the fourth year at West Point.

36. Wilson, *The Business of Civil War*, chapter 2, especially 55. Wilson cites John Dickerson, Charles Thomas, George Crosman, Stewart Van Vliet, Morris Miller,

Williams Myers, and Asher Eddy as specific West Point officers whose quartermaster general service in the West was to serve the Union well in the Civil War.

37. Scott, *Infantry Tactics; or, Rules for the Exercise and Manoeuvres of the United States Infantry 1835*. This work was published repeatedly over the next twenty-five years. Anderson, *Instruction for Field Artillery Horse and Foot*; Utley, *Frontiersmen in Blue*, 57.

38. Republished in 1863 as *Advanced-Guard, Out-Post, and Detachment Service of Troops: With the Essential Principles of Strategy and Grand Tactics for Use of the Officers of the Militia and Volunteers*.

39. Mahan, *Advanced-Guard*, 217.

40. Mahan, *Advanced-Guard*, 9.

41. Mahan, *Advanced-Guard*, 31.

42. Mahan, *Advanced-Guard*, 83–154.

43. Mahan, *Advanced-Guard*, 33–38.

44. Mahan, *Advanced-Guard*, 71.

45. Joseph Totten to Richard Delafield, Washington, 12 November 1842 and 29 November 1843, in Adjutant General War Department, USMA, 1812–67, vol. 10, 295–96 and vol. 11, 271.

46. Totten to Delafield, Washington, 29 August 1842, in Office of the Chief of Engineers, Letters from Totten 1842.

47. In Griess, "Mahan," 160; original in letter from Lt. Cyrus B. Comstock to Lt. James B. McPhearson, 18 February 1860, in James B. McPhearson Papers, MSS Division, Library of Congress, Washington DC.

48. Hope, "A Scientific Way of War," appendix E.

49. Griess "Mahan," 251; Mahan, *Summary of the Course of Permanent Fortification*, 229.

50. Griess, "Mahan," 254.

51. See Staff Library Circulation Records 1848–66, USMA Spec. Coll.

52. Griess, "Mahan," 249–86.

53. Jomini, *The Art of War*, 359; Mahan, *Summary of the Course of Permanent Fortifications*, 230.

54. The clear advantage in having several—one perpendicular—base of operations is found in Jomini's later work, but is lost among the pedantry and overemphasis of interior lines of operations. See Jomini, *The Art of War*, 1996 reprint, 77–84. The clearest articulation of advantages grated by alternate bases is in Auguste Frederic Lendy, *The Principles of War: Or Elementary Treatise On the Higher Tactics and Strategy (1853)*, a Kessinger Publishing's Legacy Reprint (Whitefish MT: Kessinger, 2014), 78-84.

55. Mahan, "Composition of Armies," 4.

56. Mahan, "Composition of Armies," 33–36. Also see letters from Mahan to Capt. George L. Welcker, 6 August and 6 September 1842, in George L. Welcker, Miscellany, Chief Engineer War Department Letters Received, 1826–66, USMA Spec. Coll.; Griess, "Mahan," 307.

57. Mahan, *Advanced-Guard*, 217.

58. USMA Staff Records, vol. 5 (1851–54), 437–54, USMA Spec. Coll.

59. See cadet notes of George Burroughs in Corps of Engineers, War Department, Substance of Lectures Delivered to the 1st Class of 1861–62 at the U.S. Military Academy by D. H. Mahan, entry 281, Burroughs Retained Papers, RG 77, NA; also see Griess, "Mahan," 238–39. For later lectures see Board of Visitor Report III (1867–68), and Lectures of Capt. Garrett J. Lydecker in George Cullum, Papers Relative to Association of Graduates, 68–72, USMA Spec. Coll.

60. Papers found in Charles Adam Dempsey Papers, USMA Spec. Coll.

61. Mordecai, *The Military Commission to Europe in 1855 and 1856*, 176.

62. Moten, *The Delafield Commission*, 180–81.

63. McClellan, *Report of the Secretary of War*, 23.

64. Delafield, *The Art of War in Europe in 1854, 1855, and 1856*, 3.

65. Richard Delafield's introductory letter to Secretary of War Jefferson Davis, in Delafield, *The Art of War in Europe in 1854, 1855, and 1856*, 1.

66. Griess, "Mahan," 214, 268–69, 272–75.

67. McClellan, *Report of the Secretary of War*, 23.

68. Griess, "Mahan," 263, 278; *Army and Navy Journal* 1 (26 December 1863): 276.

69. NASP MA, vol. 16, no. 173, January 1857, 353–60.

70. NASP MA, vol. 16, no. 179, Report by Richard Delafield to Committee on Military Affairs regarding Hermitage, 3 February 1859, 363–64.

71. Barnard, *Military Schools*, 235 (St. Cyr), 317 (Prussian cadet schools), 433–35 (Austrian Military Academies), 581–82 (Sandhurst). See also U.S. Senate, *Report of the Committee Appointed to Examine the Organization, System of Discipline, and Course of Instruction of the Military Academy*, 36th Cong., 2nd sess., Senate Misc. Doc. no. 3, 1861, 35–36 (St. Cyr), 48–49 (Metz), 61–62 (Woolwich), 66 (Prussian schools), 68 (Austrian academy), 183 (teaching of strategy), and 11 (all-arms curriculum and quotation).

72. Hope, "A Scientific Way of War," appendix D.

8. Military Science in the Civil War

1. The author has chosen to include all those graduates serving during this period, even the infirm and those forced into retirement very early in the war (albeit these were not many in number).

2. Hope, "A Scientific Way of War."

3. Wilson, *The Business of Civil War*, 72–78.

4. Figures from O. E. Hunt, "The Ordnance Department of the Federal Army," in Miller, ed., *Photographic History of the Civil War*, 5:124–25. By 1865 the ordnance bureau had overseen the manufacture of more than 1 billion rounds of rifle ammunition, and the quartermaster general had procured more than 1 million horses and mules, more than 100 million pounds of coffee, 6 million blankets, and 10 million uniforms. Also see Wilson, *The Business of Civil War*, 1.

5. Wilson, *The Business of Civil War*, 76.

6. From http://members.kos.net/sdgagnon/mil.html, accessed 27 February 2013.

7. Quotation from letters Scott wrote to Stanton, 1, 2, and 6 February 1862 (Stanton Papers, Library of Congress), found in Bruce Catton, *Terrible Swift Sword*, 147.

8. For a good analysis of Civil War strategic concepts, see Stoker, *The Grand Design*. For an excellent essay on Union thoughts on strategy, see Reardon, *With a Sword in One Hand and Jomini in the Other*.

9. From various primary sources cited in Catton, *Terrible Swift Sword*, 133.

10. For Halleck on interior lines, see *Official Record of the Rebellion*, series 1, vol. 8, 508–11. For *concentration in time*, see McPherson, *Tried by War*, 214.

11. Gay de Vernon, *Treatise on the Science of War and Fortification*, 192–93.

12. Hess, *Field Armies and Fortifications in the Civil War*, xi.

13. Catton, *Never Call Retreat*, 326–27.

14. Hess, *Field Armies and Fortifications*, xvii. See also Shiman, "Engineering Sherman's March, 60–61.

15. Editorial in *Richmond Examiner*, 23 November 1861, cited in Catton, *Terrible Swift Sword*, 107.

16. Hagerman, *The American Civil War and the Origins of Modern Warfare*.

17. *Official Record of the Rebellion*, vol. 11, part 3, 477.

18. Gabel, *Railroad Generalship*, 8.

19. The most authoritative and perhaps the best example is Sherman's *Memoirs*. See also Gabel, *Railroad Generalship*, 4–5.

20. Quoted in Royster, *The Destructive War*, 334.

Conclusion

1. Theoretically, throughout the antebellum period, the aggregate of these state forces (growing steadily from 1 million to more than 3 million in 1860) constituted sufficient military capability if required. See any *Army Register* between 1830 and 1860.

2. Maj. Gen. John Schofield, *Introductory Remarks upon the Study of the Science of War: From a Paper Read to the U.S. Military Institute, West Point, 11 October 1877* (New York: U.S. Military Service Institute, 1877), 5–11.

3. Wheeler, *A Course of Instruction*, 318.

4. Wheeler, *A Course of Instruction*, 319.

BIBLIOGRAPHY

Archival Materials

Adjutant General War Department, Records on U.S. Military Academy, 1812–67.
U.S. Military Academy Special Collection (hereafter USMA Spec. Coll.).
Mahan, Dennis Hart. "Composition of Armies," lithographic text, 1836. George L. Welcker Papers, 1841 FT 355 (2814). USMA Spec. Coll.
Manuscripts of the U.S. Military Academy, 1837–44. USMA Spec Coll.
Records of the Office of Chief of Engineers. RG 77, NA.
U.S. Congress. *American State Papers* [ASP]: *Documents, Legislative and Executive, of the Congress of the United States*. 38 vols. Washington: Gales and Seaton, 1832–61.
U.S. Military Academy Library Circulation Records, FU510.V5 UG16; Cadet Records 1824–29, 1836–60, and Staff Circulation Records 1848–66. USMA Spec. Coll.
U.S. Military Academy Staff Records. 8 vols. USMA Spec. Coll..
Bailey, Jacob Whitman. Papers. USMA Spec. Coll.
Delafield, Richard. Papers. USMA Spec. Coll.
Dempsey, Charles Adam. Papers. USMA Spec. Coll.
Engles Papers. USMA Spec. Coll.
Griswold, H. W. Papers. F. U410.D6 G889t, USMA Special Collection.
Mahan, Dennis Hart. Papers. USMA Spec. Coll.
McPhearson, James B. Papers. MSS Division, Library of Congress, Washington DC
Poinsett Papers. Historical Society of Pennsylvania and USMA Special Collection.
Swift, Joseph. Papers. USMA Spec. Coll.
West Point Thayer. Papers, 1808–72, Section 2, 25/225. USMA Spec. Coll. Two
Williams, Jonathan. Papers. Lilly Library, Indiana University.
Wood, John E. Papers. New York State Library, Albany.

Published Works

Abbot, Henry L. "Memoir of Dennis Hart Mahan, 1802–1871, by Henry L. Abbot, Read before the National Academy, Nov. 7, 1878." In *National Academy of*

BIBLIOGRAPHY

Sciences Biographical Memoirs, vol. 2. Washington: Judd and Detweiler, 1886.

Adjutant-General's Office. *Army Register.* Washington: C. Alexander, 1830–60.

Ahearn, Marie L. *Rhetoric of War Training Day, the Militia, and the Military Sermon.* Westport: Greenwood Press, 1989.

Ambrose, Stephen E. *Duty, Honor, Country: A History of West Point.* Baltimore: Johns Hopkins University Press, 1999.

American State Papers—Military Affairs. 5 vols. Washington: Gales and Seaton, 1832.

Anderson, Robert. *Instruction for Field Artillery, Horse and Foot: Translated from the French and Arranged for Service of the United States.* Philadelphia: R. P. Desilver, 1839.

Annals of Congress: Debates and Proceedings in the Congress of the United States. 42 vols. Washington: Gales and Seaton, 1832–61.

Annual Report of the Board of Visitors. West Point NY: U.S. Military Academy, 1821–46.

Annual Report of the Secretary of War, 1841. Washington: Government Printing Office, 1898.

Army and Navy Chronicles. 13 vols. Washington, 1835–42.

Barnard, Henry. *Military Schools and Courses of Instruction in the Science and Art of War in France, Prussia, Austria, Russia, Sweden, Switzerland, Sardinia, England, and the United States.* New York: Steiger, 1872.

Beirne, Francis F. *The War of 1812.* New York: E. P. Dutton, 1949.

Berard, C. *Lecon francaises a l'usage des commençants.* Paris, 1822.

Berlin, Isaiah. *Concepts and Categories: Philosophical Essays.* Edited by Henry Hardy. Princeton: Princeton University Press, 2013.

Bernardo, C. Joseph, and Eugene H. Bacon. *American Military Policy: Its Development since 1775.* Harrisburg PA: Military Service Pub., 1955.

Biot, J. B. *Élémens de la géometrié.* Paris, 1802.

Black, Jeremy. *A Military Revolution? Military Change and European Society, 1550–1800.* Basingstoke: Humanities Press, 1999.

Bland, Humphrey. *A Treatise of Military Discipline; in which is laid down and explained the duty of the officer and soldier, thro' the several branches of the service.* Boston: D. Henchman, 1743.

Bloch, Marc. *The Historian's Craft.* New York: Vintage Books, 1953.

Boorstin, Daniel J. *The Americans: The Colonial Experience.* New York: Random House, 1983.

Bristed, John. *America and Her Resources.* London: Henry Colburn, 1818.

Browning, Robert S. *Two If by Sea: The Development of American Coastal Defense Policy.* Westport CT: Greenwood Press, 1983.

Bruce, Dickson D., Jr. *Violence and Culture in the Antebellum South.* Austin: University of Texas Press, 1979.

Bülow, Heinrich, Freiherr von. *The Spirit of the Modern System of War.* Edited and translated by C. Malorti de Martemont. London: T. Egerton, 1806.

BIBLIOGRAPHY

Burr, Edward. "Historical Sketch of the Corps of Engineers, U.S. Army." *Occasional Papers, Engineers School, United States Army* 71 (Washington, 1939).

Cajori, Florian. *The Teaching and History of Mathematics in the United States.* Washington: Government Printing Office, 1890.

Cambray, Chevalier de. *Manière de fortifier de Mr. de Vauban.* Amsterdam: Pierre Mortier, 1689.

Canny, Nicholas P. "The Ideology of English Colonization: From Ireland to America." *William & Mary Quarterly* 30 (1973): 575–98.

Catalogue of Books in the Library of the Military Academy. Newburgh NY: Printed by W. M. Gazlay, 1822,

Catton, Bruce. *Never Call Retreat.* Vol. 3 of *The Centennial History of the Civil War.* Garden City NY: Doubleday, 1965.

———. *Terrible Swift Sword.* Vol. 2 of *The Centennial History of the Civil War.* Garden City NY: Doubleday, 1963.

The Centennial of the United States Military Academy at West Point, New York, 1802–1902. 2 vols. Edited by Edward S. Holden and W. L. Ostrander. Washington: Government Printing Office, 1904.

Chandler, David G. *The Campaigns of Napoleon.* New York: Macmillan, 1966.

Chorley, Katharine. *Armies and the Art of Revolution.* Boston: Beacon Press, 1973.

Clary, David A. *Fortress America: The Corps of Engineers, Hampton Roads, and United States Coastal Defense.* Charlottesville: University Press of Virginia, 1990.

Clausewitz, Carl von. *On War.* Edited and translated by Michael Howard and Peter Paret. New York: Everyman's Library, 1993.

Coffman, Edward M. *The Old Army: A Portrait of the American Army in Peacetime, 1784–1898.* New York: Oxford University Press, 1986.

Coles, Harry L. *The War of 1812.* Chicago: University of Chicago Press, 1965.

Colson, Bruno. *La Culture Strategique Americaine: L'influence de Jomini.* Paris: Economica, 1993.

Cooling, Benjamin Franklin, ed. *New American State Papers: Military Affairs.* Wilmington DE: Scholarly Resources, 1979.

Couper, William. *Claudius Crozet, Soldier-Scholar-Educator-Engineer.* Charlottesville VA: Historical Publication, 1936.

Cowley, William. "European Influences upon American Higher Education." *Education Record* 20 (1939): 172.

Cox, Jacob Dolson. *Military Reminiscences of the Civil War.* 2 vols. New York: C. Scribner's Sons, 1900. Reprint, Charleston NC: Bibliobazaar, 2008.

Crackel, Theodore J. *Mr. Jefferson's Army: Political and Social Reform of the Military Establishment, 1801–1809.* New York: New York University Press, 1987.

Cullum, George W. *Biographical Register of the Officers and Graduates of the U.S. Military Academy at West Point, N.Y., from Its Establishment, in 1804, to 1890.* 3 vols. Boston: Houghton Mifflin, 1891.

Curti, Merle E. *The Growth of American Thought.* New York: Harper & Row, 1964.

Dauben, Joseph. "Mathematics." In *From Natural Philosophy to the Sciences: Writing the History of Nineteenth-Century Science*, ed. David Cahan. Chicago: University of Chicago Press, 2003.
Dederer, John Morgan. *War in America to 1775: Before Yankee Doodle*. New York: New York University Press, 1990.
Delafield, Richard. *The Art of War in Europe in 1854, 1855, and 1856*. Washington: George W. Bowman, 1860.
Duane, William. *The American Military Library*. Philadelphia: Printed by the author, 1807–9.
Duffy, Christopher. *Fire and Stone: The Science of Fortress Warfare, 1660–1860*. Newton Abbot: David and Charles, 1975.
Dufour, Guillaume-Henri. *Cours de Tactiques*. Paris: Cherbuliez, 1851.
———. *Strategy and Tactics*. Translated by Wm. P. Craighill. New York: D. Van Nostrand, 1864.
Dupuy, R. Ernest. *Where They Have Trod: The West Point Tradition in American Life*. New York: Frederick A Stokes, 1940.
Ekirch, Arthur. *The Civilian and the Military: A History of the American Antimilitarist Tradition*. Colorado Springs: Ralph Myles, 1972.
———. *The Idea of Progress in America, 1815–1860*. New York: Columbia University Press, 1944.
Eliot, Ellsworth, Jr. *West Point in the Confederacy*. New York: G. A. Baker, 1941.
Ellicott, Andrew. *The Journal of Andrew Ellicott: Late Commissioner on Behalf of the United States . . . for Determining the Boundary between the United States and the Possessions of His Catholic Majesty in America*. Philadelphia: Budd and Bertram, 1803.
Emory, William H. *Notes of a Military Reconnaissance*. Washington: Wendell and Van Benthuysen, printers, 1848.
Endler, James R. *Other Leaders, Other Heroes: West Point's Legacy to America beyond the Field of Battle*. Westport CT: Praeger, 1998.
Falk, Stanley L. "Soldier Technologist: Major Alfred Mordecai and the Beginning of Science in the U.S. Army." PhD diss., Georgetown University, 1959.
Ferling, John E. *A Wilderness of Miseries: War and Warriors in Early America*. Westport CT: Greenwood Press, 1980.
Field, Ron. *Forts of the American Frontier, 1820–91: Central and Northern Plains*. Oxford: Osprey, 2005.
Fitzpatrick, John C., ed. *The Writings of George Washington from the Original Manuscript Sources, 1745–1799*. Vol. 26. Washington: Government Printing Office, 1944.
Flanders, Alan. "Craney Island Battle Led to Burning of Hampton." *Virginian-Pilot* (1995–10–01). http://scholar.lib.vt.edu/VA-news/VA-Pilot/issues/1995/vp951001/09290198.htm accessed 19 February 2009.
Ford, Paul Leicester, ed. *The Writings of Thomas Jefferson*. Vol. 8. New York: G. P. Putnam's Sons, 1897.

Gabel, Christopher R. *Railroad Generalship: Foundations of Civil War Strategy.* Fort Leavenworth: Combat Studies Institute, U.S. Army War College, 1997.

Gat, Azar. *A History of Military Thought: From the Enlightenment to the Cold War.* Oxford: Oxford University Press, 2001.

Gay de Vernon, Simon François, baron. *A Treatise on the Science of War and Fortification: Composed for the Use of the Imperial Polytechnick School and Military Schools and Translated by the War Department, for Use of the Military Academy of the United States: to Which Is Added a Summary of the Principles and Maxims of Grand Tactics and Operations.* 2 vols. Translated by Capt. John O'Connor. New York: J. Seymour, 1817.

General Regulations for the Army of the United States, 1841. Washington: J. and G. S. Gideon, 1841.

George, Christopher T. *Terror on the Chesapeake: The War of 1812 on the Bay.* Shippensburg PA: White Mane, 2001.

Goetzmann, William H. *Army Exploration in the American West, 1803–1863.* New Haven: Yale University Press, 1959.

Grant, Ulysses S. *Personal Memoirs of U. S. Grant.* New York: C. L. Webster, 1885.

Grenier, John. *The First Way of War: American War Making on the Frontier, 1607–1814.* New York: Cambridge University Press, 2005.

Griess, Thomas E. "Dennis Hart Mahan, West Point Professor and Advocate of Military Professionalism, 1830–1871." PhD diss., Duke University, 1968.

Guerlac, Henry. "Vauban: The Impact of Science on War." In *Makers of Modern Strategy from Machiavelli to the Nuclear Age,* ed. Peter Paret, 67–74. Princeton: Princeton University Press, 1986..

Guibert, Jacques Antoine Hippolyte de. *General Essay on Tactics: With an Introductory Discourse upon the Present State of Politics and the Military Science in Europe.* London: J. Millan, 1781.

Hagerman, Edward. *The American Civil War and the Origins of Modern Warfare.* Bloomington: Indiana University Press, 1992.

Hall, Charles B. *Military Records of General Officers of the Confederacy.* New York: Steck, 1898.

Halleck, Henry W. *Elements of Military Art and Science.* New York: D. Appleton, 1846, 1861.

Hannings, Bud. *Forts of the United States.* Jefferson NC: McFarland, 2006.

Hardee, William. *Rifle and Light Infantry Tactics.* Philadelphia: Lippincott, Grambo, 1855.

Harmon, William, and C. Hugh Holman. *A Handbook to Literature.* 7th ed. Upper Saddle River NJ: Prentice Hall, 1996.

Hart, Albert Bushnell. "The American University: Its Distinctive Feature, the 'Elective System.'" *New York Times,* 24 August 1907.

Hess, Earl J. *Field Armies and Fortifications in the Civil War.* Chapel Hill: University of North Carolina Press, 2005.

Hill, Forest G. *Roads, Rails & Waterways: The Army Engineers and Early Transportation*. Norman: University of Oklahoma Press, 1957.

Hittle, J. D., ed. *Jomini and His Summary of the Art of War*. Harrisburg PA: Military Service Pub., 1947.

Hogen, David. "Modes of Discipline: Affective Individualism and Pedagogical Reform in New England, 1820–1850." *American Journal of Education* 99, no. 1 (November 1990): 1.

Hope, Ian. "A Scientific Way of War: Antebellum Military Science, West Point, and the Origins of American Military Thought." PhD diss., Queen's University, 2012. At https://qspace.library.queensu.ca/handle/1974/7326.

Hornberger, Theodore. *Scientific Thought in the American Colleges, 1638–1800*. Austin: University of Texas Press, 1945

Horsman, Reginald. *The War of 1812*. London: Eyre and Spottiswoode, 1969.

Howe, Daniel Walker. *What Hath God Wrought: The Transformation of America, 1815–1848*. New York: Oxford University Press, 2007.

Hoyt, E. *Practical Instructions for Military Officers*. Greenfield MA: John Denio, 1811.

Hsieh, Wayne Wei-siang. *West Pointers and the Civil War: The Old Army in War and Peace*. Chapel Hill: University of North Carolina Press, 2009.

Huntington, Samuel. *The Soldier and the State: The Theory and Politics of Civil-Military Relations*. New York: Vintage Books, 1957.

Hutton, Charles. *A Course of Mathematics in Two Volumes*. New York, 1818.

Janowitz, Morris. *The Professional Soldier: A Social and Political Portrait*. Glencoe IL: Free Press, 1960.

Jeney, Louis. *Le Partisan; ou, l'art de faire la petite-guerre avec succèss selon le génie nos jours*. La Haye: H. Constapel, 1759.

Johnson, Paul E. *The Early Republic, 1789–1829*. New York: Oxford University Press, 2007.

Johnson, Timothy D. *Winfield Scott: The Quest for Military Glory*. Lawrence: University Press of Kansas, 1998.

Jomini, Antoine Henri, baron de. *The Art of War*. Translated by G. H. Mendell and W. P. Craighill. Philadelphia: J. B. Lippincott, 1862.

———. *The Art of War*. Reprint of 1862 English ed. Mechanicsburg PA: Stackpole Press, 1996.

———. *An Exposition of the First Principles of Grand Military Combinations and Movements*. Translated by J. A. Gilbert. London: T. Egerton, 1825.

———. *Summary of the Art of War: or, A New Analytical Compend of the Principal Combinations of Strategy, of Grand Tactics and of Military Policy*. Edited by Major O. F. Winship, Assistant Adjutant General, U.S.A., and Lieut. E. E. McLean, 1st Infantry, U.S.A. New York: G. P. Putnam, 1854.

———. *Treatise on Grand Military Operations*. New York: D. Van Nostrand, 1865.

Jones, J. William. *Personal Reminiscences of General Robert E. Lee*. New York: D. Appleton, 1875.

BIBLIOGRAPHY

Kaufmann, J. E., and H. W. Kaufmann. *Fortress America: The Forts That Defended America, 1600 to the Present.* Cambridge MA: Da Capo Press, 2004.

Kindred, Marilyn Anne. "The Army Officer Corps and the Arts: Artistic Patronage and Practice in America, 1820–85." PhD diss., University of Kansas, 1981.

Kite, Elizabeth. *Brigadier-General Louis Lebegue Duportail.* Baltimore: Johns Hopkins University Press, 1933.

Kohn, Richard H. *Eagle and Sword: The Beginnings of the Military Establishment in America.* New York: Free Press, 1975.

Kuhn, Thomas S. *The Structure of Scientific Revolutions.* Chicago: University of Chicago Press, 1996.

Kwasny, Mark V. *Washington's Partisan War, 1775–1783.* Kent OH: Kent State University Press, 1996.

Lacroix, M. *Élémens de l'algèbre.* Paris, 1820.

Lallemand, Henri. *A Treatise on Artillery; to which is added a summary of military reconnoitering, of fortification, of the attack and defence of places, and of castrametation.* 2 vols. Translated by Col. James Renwick. New York: C. S. Van Winkle, 1820.

Langins, Janis. *Conserving the Enlightenment: French Military Engineering from Vauban to the Revolution.* Cambridge: MIT Press, 2004.

Latimer, Jon. *1812: War with America.* Cambridge: Harvard University Press, 2007.

Leckie, Robert. *The Wars of America.* New York: Harper and Row, 1968.

Lepore, Jill. *The Name of War: King Philip's War and the Origins of American Identity.* New York: Alfred A. Knopf, 1998.

Lesage, Alain-René. *L'Histoire de Gil Blas de Santillane.* 4 vols. Paris, 1829.

Lewis, Emanuel R. *Seacoast Fortifications of the United States.* Washington: Smithsonian Institution Press, 1970.

Linn, Brian McAllister. *The Echo of Battle: The Army's Way of War.* Cambridge: Harvard University Press, 2007.

Linn, Brian McAllister, and Russell Weigley. "The American Way of War Revisited." *Journal of Military History* 66, no. 2 (April 2002): 501–34.

Lloyd, Ernest Marsh. *Vauban, Montalembert, Carnot: Engineer Studies.* London: Chapman and Hall, 1887.

Lloyd, Henry. *The History of the Late War in Germany.* 2 vols. London: S. Hooper, 1781.

Lockhart, Paul. *The Drillmaster of Valley Forge: The Baron de Steuben and the Making of the American Army.* New York: Harper, 2008.

Lynn, John A. *Battle: A History of Combat and Culture.* Boulder: Westview Press, 2003.

Macdougall, Patrick L. *The Theory of War: Illustrated by Numerous Examples from Military History.* London: Longman, Brown, Green, Longmans, and Roberts, 1858.

Mahan, Dennis Hart. *Advanced-Guard, Out-Post, and Detachment Service of Troops.* New York: E. Craighead, 1847.

———. *Advanced-Guard, Out-Post, and Detachment Service of Troops: With the Essential Principles of Strategy and Grand Tactics for Use of the Officers of the Militia and Volunteers.* New York: John Wiley, 1863.

———. *An Elementary Course in Civil Engineering.* New York: Wiley Putnam, 1837.

———. *Letter to the Honorable Mr. Hawes: In Reply to His Strictures on the Graduates of the Military Academy, by a Graduate.* New York: Wiley and Long, 1836.

———. *Summary of the Course of Permanent Fortification and of the Attack and Defence of Permanent Works, for the Use of the Cadets of the U.S. Military Academy.* 1850. Richmond: West & Johnston, 1863.

———. *A Treatise on Field Fortifications.* New York: Wiley Putnam, 1846.

Marmont, Auguste Frédéric. *The Spirit of Military Institutions.* Original 1845, reprint of 1859 ed. Translated by Frank Schaller. Columbia SC: Evans and Cogswell, 1864.

Maury, Dabney Herndon. *Recollections of a Virginian in the Mexican, Indian, and Civil Wars.* New York: Charles Scribner's Sons, 1894.

McClellan, George B. *Report of the Secretary of War Communicating the Report of Captain George B. McClellan, One of the Officers Sent to the Seat of War in Europe in 1855 and 1856.* Washington: A. O. P. Nicholson, 1857.

McDonald, Robert M. S. *Thomas Jefferson's Military Academy.* Charlottesville: University of Virginia Press, 2004.

McPherson, James M. *Tried by War: Abraham Lincoln as Commander in Chief.* New York: Penguin, 2008.

Messages of the President of the United States: with the correspondence, therewith communicated, between the Secretary of War and other officers of the government, on the subject of the Mexican War. Washington: Wendell and Van Benthuysen, printers, 1848.

"Military Academy." *North American Review* 34 (January 1832): 260.

Miller, Francis Trevelyan, ed. *Photographic History of the Civil War.* Vol. 5. New York: Thomas Yoseloff, 1959.

Millett, Allan, and Peter Maslowski. *For the Common Defense: A Military History of the United States of America.* New York: Free Press, 1994

Millis, Walter. *American Military Thought.* New York: Bobbs-Merrill, 1966.

Molloy, Peter. "Technical Education and the Young Republic: West Point as America's École Polytechnique, 1802–1833." PhD diss., Brown University, 1975.

Mook, H. T. "Training Day in New England." *New England Quarterly* 11 (December 1938): 675–97.

Moore, Jamie W. *The Fortifications Board, 1816–1828, and the Definition of National Security.* Charleston: The Citadel, 1981.

Mordecai, Alfred. *The Military Commission to Europe in 1855 and 1856: The Report of Major Alfred Mordecai, 1856.* Washington: George W. Bowman, 1860.

Morrison, James L., Jr. *"The Best School": West Point, 1833–1866.* Kent OH: Kent State University Press, 1986, 1998.

———. "Educating the Civil War Generals: West Point, 1833–1861." *Military Affairs* 38 (October 1974): 109

———. "Military Education and Strategic Thought, 1846–1861." In *Against All Enemies: Interpretations of American Military History from Colonial Times to the Present*, ed. Kenneth J. Hagan and R. Roberts William. New York: Greenwood Press, 1986.

Moten, Matthew. *The Delafield Commission and the American Military Profession.* College Station: Texas A&M University Press, 2000.

Muldoon, James. "The Indian as Irishman." *Essex Institute's Historical Collections* 111 (1975): 267–89.

Muller, Charles G. *The Darkest Days: The Washington Baltimore Campaign.* Philadelphia: J. B. Lippincott, 1963.

Muller, John. *A Treatise Containing the Elementary Part of Fortification.* London: F. Wingrave, 1766.

Muller, Julius W., ed. *Presidential Messages and State Papers.* Vol. 1. New York: Review of Reviews, 1917.

Myers, William Starr, ed. *The Mexican War Diary of General George B. McClellan.* Princeton: Princeton University Press, 1917.

Napier, Elers. *The Life and Correspondence of Admiral Sir Charles Napier.* London: Hurst and Blackett, 1862.

Nelson, Harold. "Military Roads for War and Peace, 1791–1836." *Military Affairs* 19, no. 1 (Spring 1955): 1–14.

Paine, Thomas. "Of the Comparative Powers and Expense of Ships of War, Gunboats, and Fortifications." In *The Political Works of Thomas Paine*, 2:219. London: R. Carlisle, 1819.

Pappas, George S. *To The Point: The United States Military Academy, 1802–1902.* Westport CT: Praeger, 1993.

Parker, Geoffrey. *The Military Revolution: Military Innovation and the Rise of the West, 1500–1800.* Cambridge: Cambridge University Press, 1988.

Parrington, Vernon L. *The Romantic Revolution in America, 1800–1860.* New York: Harvest Books, 1954.

Pascal, Blaise. *Pensées* and *Scientific Treatises.* Translated by R. Scofield in *The Great Books of the Western World*, vol. 33, ed. Robert Maynard Hutchins. Chicago: Encyclopedia Britannica, 1952.

Peters, Richard, ed. *Public Statutes at Large of the United States of America.* Vol. 2. Boston: Charles C. Little and James Brown, 1850.

Phillips, T. R., ed. *Roots of Strategy: The Five Greatest Military Classics of All Time.* Harrisburg: Stackpole Books, 1985.

Phisterer, Frederick. *Statistical Record of the Armies of the United States.* New York: Blue and the Gray Press, 1959.

Price, Russell R. "American Coastal Defense: The Third System of Fortification, 1816–1846." PhD diss., Mississippi State University, 1999.

Prucha, Frances Paul. *The Great Father: The United States Government and the American Indians.* Vol. 1. Lincoln: University of Nebraska Press, 1984.

Puységur, Jacques-François de Chastenet, marquis de. *Art de la guerre par principes et par règles*. Paris: Charles-Antoine Jombert, 1749.
Quartermaster's Department. *A Sketch of the Organization of the Quartermaster's Department from 1774 to 1876*. Washington: Government Printing Press, 1876.
Reardon, Carol. *With a Sword in One Hand & Jomini in the Other: The Problem of Military Thought in the Civil War North*. Chapel Hill: University of North Carolina Press, 2012.
Regulations for the Quartermaster's Department of the United States Army. Washington: J. and G. S. Gideon, 1841.
Regulations for the United States Military Academy, 1810, 1824.
Richards, George H. *Memoir of Alexander Macomb*. Bedford MS: Applewood Books, 1833.
Richardson, James D., ed., *A Compilation of the Messages and Papers of the Presidents*. Vol. 2. Washington: Joint Committee on Printing, 1908.
Riling, Joseph R. *The Art and Science of War in America: A Bibliography of American Military Imprints, 1690–1800*. Alexandria Bay NY: Museum Restoration Service, 1990.
Risch, Erna. *Quartermaster Support of the Army*. Washington: Office of the Quartermaster General, 1962.
———. *Supplying Washington's Army*. Washington: Center of Military History, U.S. Army, 1981.
Roberts, Michael. *The Military Revolution, 1560–1660*. Belfast: M. Boyd, 1956.
Robins, Benjamin. *New Principles of Gunnery*. Richmond, Surrey: Richmond Publishing, 1805.
Rochefoucault-Liancourt, Duke de la. *Travels through the United States of North America*. Vol. 2. London, 1799.
Rogers, Clifford. *The Military Revolution Debate*. Boulder: Westview Press, 1995.
Rogniat, Joseph, Baron. *Considérations sur l'art de la guerre*. Paris: Anselin et Pochard, 1820.
Rouse, Parker Jr., ed. "The British Invasion of Hampton in 1813: The Reminiscences of James Jarvis." *Virginia Magazine of History and Biography* 76 (July 1968): 318–36.
Royster, Charles. *The Destructive War*. New York: Vintage Books, 1993.
———. *A Revolutionary People at War: The Continental Army and American Character, 1775–1783*. Chapel Hill: University of North Carolina Press, 1996.
Rules and Regulations for the Government of the Military Academy at West Point. Washington: Engineer Department, 23 February 1820.
Sanders, John. *Memoirs on the Military Resources of the Valley of the Ohio, as Applicable to Operations on the Gulf of Mexico; and on a System for the Common Defence of the United States*. Washington: C. Alexander, 1845.
Saxe, Maurice de. *My Reveries upon the Art of War*. Reproduced in *Roots of Strategy: The Five Greatest Military Classics of All Time*, ed. T. R. Phillips. Harrisburg: Stackpole Books, 1985.

Schofield, John. *Forty-Six Years in the Army*. Norman: University of Oklahoma Press, 1998.

Schubert, Frank N. *Vanguard of Expansion: Army Engineers in the Trans-Mississippi West, 1819–1879*. Washington: Office of the Chief of Engineers, 1980.

Scott, Winfield *General Regulations for the Army; or, Military Institutions*. Philadelphia: M. Cary & Sons, 1821.

———. *Infantry Tactics; or, Rules for the Exercise and Manoeuvres of the United States Infantry*. New York: Harper & Bros., 1846.

Shallot, Todd Arkin. "Structures in the Stream: A History of Water, Science, and the Civil Activities of the U.S. Army Corps of Engineers, 1700–1861." PhD diss., Carnegie-Mellon University, August 1985.

Sherman, W. T. *Memoirs of General William T. Sherman*. Bloomington: Indiana University Press, 1957.

Shiman, Philip Lewis. "Engineering Sherman's March: Army Engineers and the Management of Modern War, 1862–1865." PhD diss., Duke University, 1991.

Shy, John. "American Strategy: Charles Lee and the Radical Alternative." In *A People Numerous and Armed: Reflections on the Military Struggle for American Independence*. Rev. ed. Ann Arbor: University of Michigan Press, 1990.

Skelton, William B. *An American Profession of Arms: The Army Officer Corps, 1784–1861*. Lawrence: University Press of Kansas, 1992.

Skinner, Quentin, ed. *The Return of Grand Theory in the Social Sciences*. New York: Cambridge University Press, 1985.

Smith, Gustavus W. "Company A Engineers in Mexico, 1846–1847." In *The U.S. Army Engineers—Fighting Elite*, ed. Franklin M. Davis and Thomas T. Jones. New York: Franklin Watts, 1967.

Smith, Mark Andrew. "The Corps of Engineers and National Defense in Antebellum America, 1815–1860." PhD diss., University of Alabama, 2004.

Spiller, Roger J. "John C. Calhoun as Secretary of War, 1817–1825." PhD diss., Louisiana State University, December 1977.

Starky, Armstrong. *War in the Age of the Enlightenment, 1700–1789*. Westport CT: Praeger, 2003.

Steele, Matthew. *American Campaigns*. Vol. 1. Washington: U.S. Infantry Association, 1935.

Steuben, Frederich W. A. von. *Regulations for the Order and Discipline of the Troops of the United States*. Part 1. Philadelphia: Styner and Cist, 1779.

Stevenson, Roger. *Military Instructions for Officers Detached in the Field: Containing a Scheme for Forming a Corps of a Partisan*. Philadelphia: R. Aitken, 1775.

Stoker, Donald. *The Grand Design: Strategy and the U.S. Civil War*. New York: Oxford University Press, 2010.

Strong, George Crockett. *Cadet Life at West Point*. Boston: TOHP Burnham, 1862.

Stunkel, Kenneth R. "Military Scientists of the American West." In *The U.S. Army Engineers—Fighting Elite*, ed. Franklin M. Davis and Thomas T. Jones, 61–69. New York: Franklin Watts, 1967.

BIBLIOGRAPHY

Superintendent's Curriculum Study: Report of the Working Committee on the Historical Aspects of the Curriculum for the Period 1802–1945. West Point NY: USMA, 31 July 1958. USMA Spec. Coll.

Swift, Joseph Gardner. *Memoirs of Gen. Joseph Gardner Swift LLD, U.S.A., First Graduate of the United States Military Academy, West Point, Chief Engineer U.S.A. from 1812 to 1818, 1800–1865.* To which is added a genealogy of the family of Thomas Swift of Dorchester, Mass., 1634, by Harrison Ellery. Worcester MA: F. S. Blanchard, 1890.

Thwing, Charles F. *A History of Higher Education in America.* New York: D. Appleton, 1906.

Ticknor, George. *Life, Letters, and Journal of George Ticknor.* Vol. 1. Boston: Houghton Mifflin, 1880.

Tocqueville, Alex de. *Democracy in America.* 2 vols. Edited by Phillips Bradley. New York: Alfred A. Knopf, 1945.

Totten, Joseph Gilbert. *Report of General J. G. Totten, Chief Engineer, on the Subject of National Defences.* Washington: A. Boyd Hamilton, 1 November 1851.

Traas, Adrian George. *From the Golden Gate to Mexico: The U.S. Army Topographical Engineers in the Mexican War, 1846–1848.* Washington: Office of the Corps of Engineers, 1993.

Turpin de Crissé, Lancelot, Comte. *Essai sur l'art de la guerre.* Paris: Chez Prault fils l'aîné et Jombert, 1754. Reprint, Charleston NC: BiblioBazaar, 2011.

Upton, Emory. *Military Policy of the United States.* Washington: Government Printing Office, 1911.

Vattel, Emer de. *The Law of Nations; or, Principles of the law of nature, applied to the conduct and affairs of nations and sovereigns.* Philadelphia: T. and J. W. Johnson, 1849.

Vauban, Sébastien le Prestre de. *A Manual of Siegecraft and Fortification.* Translated by George Rothrock. Ann Arbor: University of Michigan Press, 1981.

———. *Traité de l'attaque et de la défense des places fortes.* Reproduced in part in Gérard Chaliand, *Anthologie Mondiale de la Stratégie.* Paris: Robert Lafont, 1990.

———. *Le Triomphe de la Méthode: Le Traite de l'Attaque des Places.* Edited by Nicolas Faucherre and Philippe Prost. Paris: Gallimard, 1992.

Voltaire. *Histoire de Charles XII, roi de suede.* Paris: Hector Bossange, 1828.

Wade, Arthur P. "A Military Offspring of the American Philosophical Society." *Military Affairs* 38 (October 1974): 103–7.

Walker, Paul. *Engineers of Independence: A Documentary History of the Army Engineers in the American Revolution, 1775–1783.* Washington: Office of Chief of Engineers, 1981.

Watson, Peter. *Ideas: A History from Fire to Freud.* New York: HarperCollins, 2005.

Watson, Samuel J. "Knowledge, Interest, and Limits of Military Professionalism: The Discourse on American Coastal Defence, 1815–1860." *War in History* 3, no. 3 (July 1998): 280–307.

Weigley, Russell, ed. *The American Military: Readings in the History of Military in American Society.* Reading MA: Addison-Wesley, 1969.

———. *The American Way of War: A History of United States Military Strategy and Policy.* New York: Macmillan, 1973.

———. *Towards an American Army: Military Thought from Washington to Marshall.* New York: Columbia University Press, 1962.

Wesley, Edgar Bruce. *Guarding the Frontier: A Study of Frontier Defense from 1815 to 1825.* Minneapolis: University of Minnesota Press, 1935.

Wheeler, J. B. *A Course of Instruction in the Elements of the Art and Science of War for the Use of Cadets of the United States Military Academy.* New York: D. Van Nostrand, 1893.

Whisker, James B. *The Rise and Decline of the American Militia System.* Selinsgrove: Susquehanna University Press, 1999.

White, Leonard. *The Jacksonians: A Study in Administrative History 1829–1861.* New York: Free Press, 1954.

Wilcox, Gen. C. M. *History of the Mexican War.* Washington: Church News, 1892.

Williams, T. Harry. *Americans at War: The Development of the American Military System.* Baton Rouge: Louisiana State University Press, 1960.

———. *With Beauregard in Mexico: The Mexican War Reminiscences of Beauregard.* Baton Rouge: Louisiana State University Press, 1956.

Wilson, Mark R. *The Business of Civil War: Military Mobilization and the State, 1861–1865.* Baltimore: Johns Hopkins University Press, 2006.

Winders, Richard. *Mr. Polk's Army.* College Station: Texas A&M University Press, 1997.

Woodward, Augustus Brevoort. *A System of Universal Science.* Philadelphia: Edward Earle, Harrison Hall, and Moses Thomas, 1816.

Wyatt-Brown, Bertram. *Southern Honor: Ethics and Behavior in the Old South.* New York: Oxford University Press, 2007.

INDEX

Page numbers in italic indicate illustrations.

Abert, John James, 116, 193, 304n44
Academic Board, 61–63, 67, 73, 77–79, 85, 122, 146–47, 152, 159, 298n40
active defense, 101, 227
Adams, John, 27, 34, 282–83n44
Adams, John Quincy, 283n53
Adolphus, Gustavus, 84
Advanced-Guard, Out-Post, and Detachment Service of Troops (Mahan), *155*, 195–99, 201–2, *202*
American Artillerist's Companion (Tousard), 82
The American Military Library (Duane), 39–40
American Philosophical Society, 36
American Revolution, 17–26, 45, 245, 281n9, 290n11
The American Way of War (Weigley), 9
Anderson, Robert, 195, 209
antebellum military science, 183–212; Third System criticism of, 184–93, *191*; at USMA, 183, 190, 195, 197, 200–211; and westward expansion, 193–200
antiprofessionalism, army, 130–35
Armstrong, John, 44
Army of Northern Virginia, 1, 221, 224, 229, 242
Army of the Cumberland, 221
Army of the Ohio, 221

Army of the Potomac, 1, 221, 223, 228, 231
Army of the Tennessee, 221, 241
army professionalization, 9, 47, 246, 249
army reforms, 1815–1820, 47–75; formation of staff bureaus, 47, 51–53, 62–63, *63*, 74, 248; officer corps professionalization, 47; professionalization, 9, 47, 246, 249, 254; Thayer system of USMA cadet selection, 64–68; and Third System defenses, 51–59, *58*, 285n12; and USMA reforms, 59–63; and USMA scientific curriculum, 68–74; and War of 1812, 48–51, 284–85n4
artillery: in Civil War, 216, 223, 230, *234–35*, 244; Duane on, 40–41, 44; and mathematics, 3; and Third System reforms, 57, 61–63; in USMA curriculum, 72, 74, 78–90, 92–97, 103–5
art of war, 41, 140, 206, 252, 278n12

Bailey, Jacob W., 146
Baltimore & Ohio Railroad, 225
Banard, John, 146
Barnard, Henry, 77
Barriffe, William, 19
Bartlett, William H. C., 146
bases of operation: in Civil War, 213, 222–28, *222*, 231–33, *233*, 238–53, *239*, *241*, 251; Duane on, 40–41; Halleck on, 185; importance of, 1–6; Jomini on, 98,

INDEX

bases of operation (*cont.*)
  308n54; Mahan on, 153–58, 200–203, *202*;
  in Mexican War, 162, 164, 166–70, 302n2;
  O'Connor on, 101–2; in USMA curriculum, 98, 101–5; von Bülow on, 41–43, *42*;
  Wheeler on, 252
bases of supply, 187, 222. *See also* bases of operation
bastions, 93–97, 150
Beauregard, Pierre G. T., 11, 168, 170, 225, 228
Bérard, Claudius, 60, 146
Berlin, Isaiah, 279n19, 298–99n45
Bernard, Simon, 54–56, 62–63, 72–73, 112, 116, *120*, 131
Black, James A., 142, 145
Black Hawk War, 133, 135, 187
Blake, Jacob, 163
Bland, Humphrey, 19, 24
Bliss, William W. S., 163–64
Bloch, Marc, 10
Blue Book (Steuben), 23, 31, 52
Board of Engineers for Internal Improvements, 107, 112–13, 126, 138
Board of Visitors, 60–62, 72, 81, 121, 140–43, 148, 159, 206, 295n3, 299n47, 299n61
Bomford, George, 163
Bonaparte, Napoleon, 7–8, 40, 54, 81, 90–91, 99, 102, 179, 182, 196–97
Bragg, Braxton, 11, 165, 168
Brewerton, Henry, 146
Bristed, John, 288n47
Brown, Jacob, 165
Brown, John, 210
Buell, Carlos, 225–26, 228, 230, 232
Bülow, Heinrich Dietrich von, 41–43, *42*, 91, 98, 101–2, 153–54, 157, 185, 200, 284n71
Burbeck, Henry, 34–35
Burnside, Ambrose, 11, 228

Calhoun, John C., 53, 56–59, 63, 70–74, 104, 109, 131, 163, 209, 246, 286–87n30
campaign theory, 5–6, 40–41, 88–89, 92–93, 99, 105, 149–54, 157–59, 179–80, 213
canal building, 109–12

Carnot, Lazare, 111
cavalry, 78–79, 85, 129, 136, 147, 189, 207–9, 297n26
ceinture de fer (iron belt), 4, 11, 26, 55
celerity of movement, 44, 101, 105, 155–56, 170, 196, 214, 228
Centreville VA, 223, 225, 229, *234*
Charles, Archduke of Austria, 43, 102, 153, *155*, 157, 185, 197, 200, 307n23
Charles XII, king of France, 84
Chase, W. H., 184, 188, 192, 305–6n2
Chassuers Britannique, 284–85n4
Chesapeake incident, 38
Church, Albert E., 146, 164
citizen militias. *See* militias, citizen
City Point wharf, *235*
civil engineering: in Civil War, 231; colonial era, 23; Mahan on, 122–25, 129, 148, 159; in USMA curriculum, 12, 83, 90, 107, 113–16, 121–27, 148, 159, 203. *See also* military engineering
Civil War: artillery in, 216, 223, 230, *234–36*, 244; civil engineering in, 231; fortifications in, 213–16, 221–24, 227, 229–30, *234–35*, 250; as last Napoleonic war or first modern war, 253; logistics in, 213, 215, 218, 225, 231–32, 238, 242, 244, 250–51; military departments in, *220*; military science in, 1–2, 213–44, 251; mobilizing and organizing for, 216–43, *220*, *241*, 309n4; siege craft in, 1–2, 213, 216, 223, 233, 238, 250; strategy in, 213, 220–21, 226–28, 243, 250; Third System defenses in, 212–13, *213*, 216, *222*, 223, 227; and USMA, 213–16, 218–19, 223–24, 227–31, 243, 309n1
Clausewitz, Carl von, 9, 69, 98, 252–53, 307n23
coastal fortifications, 11, 17, 31–33, 45, 47–50, 53–57, *120*, 131–32, 190–93, 245–46, 279n26, 282n43, 286n21, 286n25
Cockburn, George, 48
Coffman, Edward, 293n37
The Compleat Body of the Art Military (Elton), 19
concentration of force, 101–5, 154, 214, 227

conscience, defined, 5
conscription, 50, 213, 217
constitution of 1789, 30
Continental Army, 17, 23, 25, 29, 245
corps of engineers, 33–35, 39, 45, 50, 57, 110, 123, 125, 131, 136
coup d'œil, 8, 71, 89, 92, 116, 171, 278n17. *See also* genius, military
Cours de Tactique (Dufour), 201
A Course of Instruction in the Elements of the Art and Science of War (Wheeler), 251–52, *252*
Cram, Thomas, 163
Crawford, William H., 55, 60, 109
Crimean War, 205, 207
Crogan, George, 304n44
Cross, T., 119
Crozet, Claudius, 60–61, 73, 83, 113

Dade, Francis Langhorne, 133
Davis, Jefferson, 174, 193, 204, 208–9, 217, 304n45
Davis Commission, 14, 209
Deane, Silas, 23
de Cormontaigne, Louis de, 92, 197
Delafield, Richard, 146, 148, 159, 197, 204–8
Delafield Commission, 14, 206
de la Rochefoucault-Liancourt, Duke, 32
Dempsey, Charles, 203–4
depots, 29–33, 42, 85, 88, 98–99, 104, 154, 164–65, 194, 221–22, *222*
descriptive geometry, 83
dialectics, 68, 247
doctrine, 9–12, 14–16, 277n6
Doniphan, Alexander, 162
Douglas, David, 113
Duane, William, 39–44, 284n70
Dufour, Guillaume-Henri, 186, 201, 306n9
Duportail, Louis Lebègue, 23, 25–27, 31, 53

Eastern Department, 132, 138
Eaton, Amos, 113
École du Corps Royale du Genie (France), 3, 23
École Militaire (France), 3

École Polytechnique (France), 3–4, 37, 65–66
An Elementary Course on Civil Engineering (Mahan), 123
Elements of Military Art and Science (Halleck), 185–90, 196
Ellicott, Andrew, 60
Elliot, J. D., *120*
Elton, Richard, 19
Emory, William H., 137, 283n62, 297n29
empiricism, 69, 84
engineer-artillerists, 25–26
engineering. *See* civil engineering; field engineering; military engineering
England: boundary disputes with, 137–38; costal defenses against, 132; military methods of, 19, 70–71; mobilization needs of, 188; romantic view of military genius in, 278n18; threat of war with, 31, 35, 38
Enlightenment: and Bristed, 288n47; and Jomini, 9; and liberal arts, 178–79; and mathematics, 4, 283n58; and military science, 4–11, 70, 86, 91, 105, 157, 199, 209, 211, 215, 247, 254, 278–79n18; and romanticism, 141; and U.S. Military Philosophical Society, 36–37; von Bülow on, 284n71
entrenchment. *See* trench warfare
Erskine, Robert, 23
Essai sur l'art de la guerre (de Crisse), 40

Farragut, David, 227
field engineering, 14, 148, 150, 214, 228–31, 239
First System defenses, 30–38, *32*
First World War, 253
foraging, 88, 158, 186, 231, 238
force, concentration of, 101–5, 154, 214, 227
Fort Brown, 165, 167, 303n16
fortifications: antebellum era, 183–92, *191*, 196–200, 204–12; in Civil War, 213–16, 221–24, 227, 229–30, *234–35*, 250; coastal, 11, 17, 31–33, 45, 47–50, 53–57, 131–32, 190–93, 245–46, 279n26, 282n43; colonial era, 17–21, 24, 26; de Vauban on, 4–5, 13, 24, 26, 33–35, 42, 55, 92–94, *94*, 96–99, 150–51, 227; early national, 29–30; field,

INDEX

fortifications (*cont.*) 96; First System defenses, 31–38, *32*; Gaines on, 297–98n34; Halleck on, 186, 278n12; Jacksonian era, 129, 132–33, 138–39, 143, 148–52, 158–60; Mahan on, 13, 148–52, 158–59, 196–200, 206–7, 224, 300nn66–67; and mathematics, 3; in Mexican War, 165–67, 172, 181; Partridge on, 297–98n34; permanent, 96; and Third System reforms, 53–59; in USMA curriculum, 82–83, 86–87, 90–94, *94*, 96–98, *100*, 104–5; Wheeler on, 252

Fortifications Board, 55–59, 62, 111, 115–16, 132, 207, 286n25

Fort Taylor, 303n16

Fort Texas, 165–66, 303n16

Forty-Six Years in the Army (Schofield), 293n37

France: and the American Revolution, 23–25; mobilization needs of, 188; and U.S. military reforms, 53–55

Franklin, Benjamin, 23, 36

Frederick the Great, 8

French and Indian War, 20

Fries's Rebellion, 33

frontier military roads, 118–21, *120*

Fulton, Robert, 283n53

Gaines, Edmund P., 133, 135–36, 138–39, 142, 149, 173, 184, 188, 287–88n30, 297–98n34

Gaines, Edward, 58, 102, 287n30

Galvin, John, 10

de Vernon, Simon François, Gay, 86, 91–94, *94*, 96–99, *100*, 103–4, 137, 151–53, 156, 198, 291n41, 291n44

Geist des neuern Kriegssystem (Bülow), 41

General Regulations for the Army (Scott), 52, 82

general staff system. *See* staff bureaus

General Survey Act of 1824, 112

genius, military: Antoine Jomini on, 189; British perceptions of, 70–71; changing perceptions of, 130, 141, 278–79n18; Calhoun on, 70; Confederate perceptions of, 243; conflicting perceptions of, 7–9; de Vernon on, 93, 103; English view of, 278n18; Lallemand on, 88; Mahan on, 179; Napoleon as, 179, 182; O'Connor on, 104; in state militias, 15–16; Union advantage of, 243–44

geometry, 3–4, 7–8, 35, 40, 43, 82–83, 98–99, 200

Gimbrede, Thomas, 84

Grant, Ulysses S.: academic prowess of, 164, 302n13, 303n14; as assistant quartermaster, 194; and bases of operations, 201; and Civil War, 23, 221, 225–26, 228, 230–33, 239–43, *241*; and mathematics, 164, 303n14; and Mexican War, 164, 171; and military science, 1, 11, 14, 277n1; as realist, 250

Gratiot, Charles, 131

Great Military Road, 119, 135

Green, Nathanael, 24

Greenville Treaty, 32

Grenier, John, 21–22, 281n14

Gridley, Richard, 23, 281n9

Griess, Thomas, 124, 289n9

Guibert, Jacques Antoine Hippolyte de, 91, 278–79n18

Halleck, Henry, 11, 103, 174, 184–90, 196, 209, 214, 226, 228, 278n12, 306n5, 306n19, 307n20, 307nn22–23

Hamilton, Alexander, 26, 29, 33, 50, 282n44, 287–88n30

Hardee, William, 195, 209

Harpers Ferry Armory, 163, 210

Harvard University, 114

Haupt, Herman, 224, 230

Hawkins, Edgar, 165

Heath, William, 287–88n30

Hess, Earl J., 229

History of Charles XII (Voltaire), 84

Hittle, J. D., 156

Hood, John Bell, 11, 240–41

Hooker, Joseph, 11, 228

Hoyt, Epaphras, 43–44

Huntington, Samuel, 10, 85, 114–15, 118, 142

Indians Removal Act of 1830, 133–34

INDEX

Indian wars, 132–36, 187, 249
individualism, 61, 68
Infantry and Cavalry School (France), 65
Infantry Tactics (Scott), 82, 195–96, 209
interior communications, 57, 109
internal military improvements, 107–27; frontier military roads, 118–21, *120*; shift to, 107–18, 292n10; USMA civil engineering in context, 121–26
irregular warfare. See *la petite guerre*

Jackson, Andrew: and cadet selection at USMA, 64; and criticism of staff bureaus, 173; and frontier military roads, 109; and Indians Removal Act of 1830, 133–34; and Mahan, 130, 295n5; and Mexican War, 173; and partisan warfare, 22; and professionalism, 130–31; support of USMA, 130, 295n5; and War of 1812, 51. See also Jacksonian military science
Jackson, Thomas, 11, 228, 233
Jacksonian military science, 129–60; overview, 129–35; shift from Atlantic to western focus, 135–39; and state militias, 134, 296n18; and Third System defenses, 296n12; and USMA, criticism and defense of, 140–46; and USMA curriculum changes, 146–60, 301–2n94, 301nn87–88; and War of 1812, 132, 295–96n11
James River, 48, 231, *237*, 240
Jay Treaty, 32
Jefferson, Thomas, 34–36, 38, 64, 108, 110, 127, 247, 283n53
Jesup, Thomas, 52, 162, 194, 304n44
Johnston, Albert Sydney, 10, 225–26, 228, 230, 240, 242
Jomini, Antoine Henri, Baron de: on bases of operation, 201, 308n54; campaign theory of, 185; and Duane, 40; and Enlightenment military reasoning, 9; on geographic theater of war, 96; and geometry of military movements, 43; on lines of operation, 101–2, 153, 156; and Mahan, differences between, 153; on military genius, 189; on military science, 7; publications of, 156; reputation of, 156

Jones, Roger, 162, 173–74, 304n44

Kearny, Stephen, 137, 162
King George's War, 20
King William's War, 20
Knowles, Robert, 138
Knowlton, Miner, 195
Knox, Henry, 22, 25–27
Knox, John, 30–31
Kuhn, Thomas, 10

Lafayette, Marquis de, 23
la grande guerre: in antebellum era, 181, 195, 197, 210; in Civil War, 216, 243; defined, 12; Delafield on, 206; de Vernon on, 92; Mahan on, 122; and Topogs, 110; in USMA curriculum, 105, 107, 142, 246
Lallemand, Henri, 86–90, 137, 151, 291n36
la petite guerre, 15, 17, 21–22, 92, 195–96, 201, 243, 253, 280n29
Lawson, Thomas, 304n44
Lee, Charles, 22
Lee, J. E., 304n44
Lee, Robert E.: as Academic Board superintendent, 146; and Civil War, 221, 228–29, 238, 240, 242, 250; and Delafield Commission court of inquiry, 207; and Mexican War, 166, 168, 170; and military science, 1–2, 10, 15
Lendy, Auguste Frederic, 186, 201
L'Enfant, Pierre Charles, 26–27, 31
les Lumières, 4, 8, 18, 37, 98, 284n71
levée en mass, 15. See also citizen militias
Lévy-Buhl, 10
Lewis, Meriwether, 110
liberal arts: current military perspective of, 8; in USMA curriculum, 65, 73, 140–42, 158, 178, 247
Lincoln, Abraham, 88, 180, 217, 224, 226, 243, 277n1
lines of operation: in Civil War, 225–28, 232–33, *233*, *239*; early theory of, 41–43, *42*; Halleck on, 185–88; Jomini on, 101–2, 153, 156; Mahan on, 153–57, *155*, 200–201; in Mexican War, 162, 166–68; and Scott, 102; in USMA curriculum, 99–103; Wheeler on, 252

329

Linn, Brian McAllister, 9–10
living off the land. *See* foraging
Lloyd, Henry, 40–41, 91, 101, 278–79n18
logistics: in Civil War, 213, 215, 218, 225, 231–32, 238, 242, 244, 250–51; Duane on, 41; Halleck on, 186–87, 278n12; and internal improvements, 115, 118, 127; ; Mahan on, 201–2; in USMA curriculum, 87–89, 105
Long, Stephen H., 110
Louisville & Nashville Railroad, 225
Louis XIV, king of France, 3–4

Macdougal, Patrick, 185, 305n8
Macomb, Alexander, 72, 109, *120*, 131, 173, 286n21
Madison, James, 47, 53–54, 74, 283n53, 285n12
Mahan, Dennis Hart: advanced engineering program, 215, *271–74*; on alternative military options, 16; and antebellum USMA curriculum, 195–203, 207; on bases of operation, 102–3, 153–57, *155*; on citizen militias, 300n75; and civil engineering instruction, 122–25, 129, 148, 159; and Civil War, 214–15, 223–24, 229; death of, 251; and Delafield Commission Reports, 206–7; on fortifications, 13, 148–52, 158–59, 196–200, 206–7, 224, 300nn66–67; on French vs. English military tactics, 16; and Dufour, 306n9; as head of Academic Board, 146; honorary degrees of, 294n47; and Jackson, 130, 295n5; and Jacksonian era USMA curriculum, 129, 140, 146, 148–59, *155*, 301–2n94, 301nn87–88; and Jomini, differences between, 153, 300–301n79; on lines of operation, 153–57, *155*; on logistics, 201–2; on mathematics of range estimation, 300n66; on military engineering, 148, 152, 158; on military genius, 179; nickname of, 295n53; as president of Napoleon Club, 181; on strategy, 152–61, *155*, 197–204, *202*; on tactics, 140, 197–99; on topographical engineers, 151, 300n73
Manassas Gap Railroad, 225

Mansfield, Jared, 60
Mansfield, Joseph, 165
Marcy, William, 162, 172
Marmont, Marshall, 185, 305n8
Martemont, C. Malorti de, 71, 279n18
mathematics: and artillery, 3; de Vauban on, 4–5; and Enlightenment, 4, 283n58; and fortifications, 3; and Grant, 164, 303n14; and military science, 3–4, 254; Mahan on, 300n66; in U.S. colleges in nineteenth century, 299n46; in USMA curriculum, 65, 72–74, 80–83, 87, 89, 96, 140–42, 247, 254, 283n58
McClellan, George, 11, 167–68, 180, 204–7, 223, 225–26, 228–30, 238
McDowell, Irwin, 10, 225
McHenry, James, 26
McHenry, William, 33–34, 282–83n44
McRee, William, 54–55
Meade, George, 1–2, 11, 163, 166
Meigs, Montgomery, 218–19, 224, 241
Memphis & Charleston Railroads, 226
the *Merrimac*, 192
metaphysics, 8, 68–70, 160
Mexican War, 161–82; aftermath of, 172–81; army strength during, 131; citizen militias in, 162, 302n5; policy and strategy of, 161–68, *165*, 171, 175, 181; and USMA, 161–64, 166–68, 171–73, 177–82, 249; Vera Cruz to Mexico City, 168–72
Military Academy at Wiener-Neustadt (Austria), 3
Military Discipline (Barriffe), 19
military engineering: in frontier service, 122; Hoyt on, 44; Mahan on, 148, 152, 158, 199; in Mexican War, 181; in USMA curriculum, 86, 90, 105, 121, 140–42, 203
military genius. *See* genius, military
Military Instructions for Officers Detached in the Field (Stevenson), 19, 43
military journals, 204
military policy: first U.S. national, 74; Halleck on, 187; Jacksonian era, 142; Weigley on, 9
military reforms, 1815–1820. *See* army reforms, 1815–1820

INDEX

military roads, frontier, 118–21, *120*
military science: Calhoun on, 53, 59, 63; in Civil War, 1–2, 213–44, 251; colonial era, 17–26; defined, 5; early national, 26–30; English view of, 70; and Enlightenment, 4–11, 70, 86, 91, 105, 157, 199, 209, 211, 215, 247, 254, 278–79n18; First System defenses, 30–38; Jomini on, 7; and mathematics, 3–4; and militia officers, 39–44; origins of, 2–3; overview, 1–16; as reflection of social trends, 7–8; and Lee, 1–2, 10, 15; and Scott, 10; Second System defenses, 38–39; and staff bureaus, 52; and Thayer, 54; and Totten, 10; USMA scientific curriculum 1815–1820, 68–74; USMA scientific curriculum ca. 1825, 77–105, 289n9. *See also* antebellum military science; Jacksonian military science
militia officers, 39–44
militias, citizen: as alternative military option, 15; antebellum era, 187–88, 191–92, 197–98, 246, 310n1; Calhoun on, 72; Cannon on, 70; colonial era, 21–22, 24, 27–33, 281n13; early attitudes toward, 70, 246; and internal improvements, 115, 118; Jacksonian era, 129–36, 145, 149–52, 160; Mahan on, 151–52, 300n75; in Mexican War, 162, 302n5; military genius in, 15–16; Organization of the Militia Act of 1790, 30; and Third System reforms, 51, 53, 55–59, 74; Uniformed Militia Act of 1792, 31; in War of 1812, 48–50
Mobile & Ohio Railroads, 226
the *Monitor*, 192
Monroe, James, 50–56
Monroe, James, 47, 74, 246
Mordecai, Alfred, 204–5
Morrison, James L., 295n58, 301n84
Moten, Matthew, 142
Motier, Gilbert du (Marquis de Lafayette), 23
Muller, John, 19
Munroe, John, 164

Napoleon Club, 13, 179–82, 198, 215, *261–70*, 305n60

Napoleonic Wars, 35, 54, 179, 185
New Principles of Gunnery (Robins), 19
Northeast Boundary Survey, 137
Norton, James St. Clair, 184
Norwich University, 65

objective points, 41–42, *42*, 99, 153–54, 166–68, 200, 242, 252
O'Connor, John Michael, 91–94, 98–105, *100*, 152–53, 155–56
Official Record of the Rebellion, 232
The Old Army (Coffman), 293n37
On War (Clausewitz), 69, 307n23
Operation Desert Storm, 253
Orange & Alexandria Railroad, 225, 230
Organization of the Militia Act of 1790, 30

parapets, 31, 95
partisan warfare, 22
Partridge, Alden, 60–61, 65, 297n34
Pascal, Blaise, 7–8
Patterson, Robert, 167
Pemberton, John, 11
Permanent Fortifications (Mahan), 152
personal honor, 61, 68, 287n35
Petersburg VA, 2, 14, 216, 224, *235*
Pickering, Thomas, 24
Pike, Zebulon, 110
Poinsett, Joel, 134–38
Polk, James, 161–62, 166–67, 174
Pope, John, 228, 230
Practical Instructions for Military Officers (Hoyt), 43
Principles of War (Lendy), 201
professionalization, army, 9, 47, 246, 249, 254
Putnam, Rufus, 23

quartermaster general, 51–52, 116–17, 119, 121, 126, 176, 185, 194, 218–19, 307–8n36, 309n4
Quasi-War of 1798, 32–33
Queen Anne's War, 20

railroads, 113, 124, 139, 160, 191–94, 210, 219, 224–25, 232, 250, 307n35
ramparts, 94–95, 207, 223, 229
realism, 11, 250

331

reconnoitering, 87, 89, 111, 291n36
"Reduction of the Army" (Calhoun), 58–59
Regulations for the Order and Discipline of the Troops of the United States (Steuben), 23, 31, 52, 82
Rensselaer, Stephen van, 113
Rensselaer School, 65, 113–14
Renwick, James, 86–87
"Report on Fortifications" (1821), 109
"Report on Plans for Protection of Northern and Eastern Boundary" (Poinsett), 138
"Reports on Roads and Canals" (Calhoun), 109
Reynolds, John, 133
Rifle and Light Infantry Tactics (Hardee), 195
Ringgold, Samuel, 195
Roberdeau, Isaac, 112
Robins, Benjamin, 19
Rochefontaine, Stephen, 33–34, 282n44
romanticism, 11, 140–41, 160, 178, 182, 250–51, 278n18, 279n19, 288n47, 298–99n45
Rosecrans, William, 228, 238–39
Royal Artillery Depot (England), 3
Royal Military College (England), 65
Russy, Réne de, 146

Sanders, John, 164, 184
Saxe, Maurice de, 25, 278–79n18
Schofield, John, 251, 293n37
science, defined, 4–5
science of commanders, 40. *See also* strategy
science of movements, 5–6, 41, 154, 157, 181, 278n11. *See also* strategy
science of war: Board of Visitors on, 141; defined, 5; ; Duane on, 40, 43; English view of, 71; and France, 81; Gaines on, 136; de Vernon on, 86, 91; Mahan's course on, 152, 158–59, 166, 197, 215; and Napoleon Club, 179; O'Connor on, 101–2; USMA commitment to, 251–52. *See also* military science

Scott, Winfield: and Board of Visitors recommendations, 141–42, 159, 295n3; and Civil War, 225–26, 228; as Eastern Department commander, 58, 131–32, 135, 138; and Gaines, 135; and lines of operation, 102; and Mexican War, 162–63, 167–74, 178; and military science, 10; publication of *Infantry Tactics*, 195–96, 209; and science of movements, 278n11; and staff bureau regulations, 52; and Third System defenses, 131
Second System defenses, 38–39, *39*, 48–49
Second World War, 253
sectionalism, 207
seizing the initiative, 101, 155–56
Seminole War, 133, 135, 249
Senate Committee on Retrenchment, 142
"Sentiments upon a Peace Establishment" (Washington), 27, 108, 287–88n30
Sheridan, Philip, 240
Sherman, William T., 1–2, 11, 14, 201, 221, 229, 231–32, 239–43, 251
siege craft: in Civil War, 1–2, 213, 216, 223, 233, 238, 250; colonial era, 20, 22, 24; de Vauban on, 4, 24, 40, 151, 201; Mahan on, 151, 154; and mathematics, 4; in Mexican War, 162, 168–69; in USMA curriculum, 86, 93, 96–99, 181
siege of Sebastopol, 205, 207
siege parallels, 24, 40, 93, 99
Skelton, William B., 142, 287n32, 293n37
Smith, Gustavus, 167–68, 180, 303n22, 303n24
social respectability, 118, 293n37
Sparks, Richard, 110
Springfield Armory, 163
Staff Act of 1818, 51–52
staff bureaus: in Civil War, 216; difficulties with, 304n44, 304n50, 305n53; formation of, 47, 51–53, 62–63, *63*, 74, 248; and internal improvements, 107, 116–18; in Mexican War, 161; after Mexican War, 172–77; USMA graduates in, 248–49, 304–5n52
state-on-state warfare. See *la grande guerre*
states' rights, 53, 115, 118, 130, 218, 243

Staunton-Parkersburg Pike, 225
St. Clair disaster of 1791, 31
steam technology, 12, 129, 138–39, 147, 190, 192
Steuben, Friedrich Wilhelm von, 23, 26, 29, 31, 52, 82, 281n20
Stevenson, Roger, 19, 43
Stoneman, George, 240
strategic points, 6, 153–54, 156, 186, 191, 200, 213, 219, 224–29, 250
strategic thinking, 120, 190, 295n58
strategy: antebellum era, 183–85, 197–204, *202*, 208–10; in Civil War, 213, 220–21, 226–28, 243, 250; defined, 6–7; Duane on, 40, 284n70; Halleck on, 185, 278n12; Hoyt on, 44; Lallemand on, 89; Mahan on, 148, 152–61, *155*, 197–204, *202*; during and after Mexican War, 161–68, *165*, 171, 175, 181; in USMA curriculum, 87–91; von Bülow on, 41
The Structure of Scientific Revolutions (Kuhn), 10
"On the Subject of National Defenses" (Totten), 190
sublime qualities, 8, 71, 89, 202, 214, 278n17, 278n18. *See also* genius, military
Sullivan, John L., 112
Summary on the Art of War (Jomini), 156
Swift, Alexander J., 167, 303n22
Swift, Joseph G., 53–54, 63, 72, 81, 86

tactics: antebellum era, 195–200, 208, 210; defined, 6–7; Halleck on, 278n12; Hoyt on, 44; Jacksonian era, 140, 147, 150, 153, 160; Mahan on, 140, 197–99; in USMA curriculum, 80, 82–83, 86, 98, 103–5; von Bülow on, 41
Talcott, George, 304n44
Taylor, Zachary, 162–68, 172–73, 303n16, 303n18, 303n24
Texas Boundary Survey, 137
Thayer, Sylvanus, 47, 49, 54, 60–61, 63–69, 73–75, 146–47
Thayer system of USMA cadet selection, 64–68
theater of war, 6, 98–99, *100*, 102, 153, 168, 185–86

Third System defenses: antebellum era criticism of, 184–93, *191*; in Civil War, 213, 216, 221–23, *222*, 227, 250; creation of, 51–59, *58*, 285n12; and Delafield Commission court of inquiry, 207; and internal improvements, 108, 111–12, 115–16, 125–26; Jacksonian era, 129–42, 160; Mahan on, 199; McClellan on, 207; in Mexican War, 162, 167; naming of, 280–81n26; publication of, 298n37; in USMA curriculum, 73, 88, 90, 94, 199
Thomas, Pierre, 60
Ticknor, George, 68–69
Tocqueville, Alex de, 295n1
topographical engineers (Topogs): Abert as chief, 302n10; in Civil War, 228, 231; formation of, 110–11, 126; and internal improvements, 114, 117; Jacksonian era, 136–38, 297n27; Mahan on, 151, 300n73; and Mexican War, 163, 166, 168, 171; surveying western frontier, 137–38; and westward expansion, 193–94
topography: in USMA curriculum, 89, 92–93, 96, 99, 103; Vauban on, 5–6
Totten, Joseph: and antebellum USMA curriculum, 197–98; on Board of Engineers, 112, 116; and Board of Visitors recommendations, 159; on Fortifications Board, 55; and Mexican War, 168; and military science, 10; report on national defenses (1851), 190–93; and Third System defenses, 131, 184, 190–93; and USMA criticism and defense of, 142–48, 294n51, 299n54, 299n61, 304n44
Tousard, Louis de, 82
Tower, Zealous, 170
A Treatise Containing the Elementary Part of Fortification (Muller), 19
A Treatise of Military Discipline (Bland), 19, 24
Treatise on Artillery (Lallemand), 86–90
Treatise on Field Fortifications (Mahan), 149–56, *155*
Treatise on Grand Operations (Jomini), 156

Treatise on the Science of War and Fortifications (de Vernon), 86–87, 91–96, *94*, 291n41
trench warfare: in American Revolution, 22; in Civil War, 1, 223, 229–30, 240, 250; Mahan on, 149–51; in Mexican War, 169, 174, 181; in USMA curriculum, 84, 87, 96–97; in World War I, 253
trigonometry, 82–83, 143, 151
Trist, Nicholas, 64
Turpin de Crissé, Lancelot, 25, 40, 278n18

Uniformed Militia Act of 1792, 282n39
universal military service, 20–21
Upton, Emory, 283n62
U.S. Corps of Engineers. *See* corps of engineers
U.S. Military Academy (USMA): in antebellum era, 183, 190, 195, 197, 200–212, 247, 280n27; appropriations for, 130, 295n3; cadet routine ca. 1825, 77–80, 289nn3–4; cadet selection, 64–68; and Civil War, 213–16, 218–19, 223–24, 227–31, 243, *275*; class divisions, 66; criticism and defense of, 140–46; curriculum after War of 1812, 65–74, 288n52; domination of staff bureaus, 248–49; establishment of, 33–39, 246; First Class instruction ca. 1825, 90–104, 291n41, 291n44; Fourth and Third Class instruction ca. 1825, 81–84, 290n11; French language instruction, 80–81, 86, 135–36, 147; Jacksonian era curriculum changes, 146–60, 301–2n94, 301nn87–88; mathematics instruction at, 65, 72–74, 80–83, 87, 89, 96, 140–42, 247, 254, 283n58; and Mexican War, 161–64, 166–68, 171–73, 177–82, 249; Napoleon Club, 13, 179–82, 198, 215, *261–70*; Newton Cannon on, 70; novels allowed at the library, 178–79; officers performing staff functions, *259–60*; reforms 1815–1820, 59–63; regulations, 1810, 37; regulations, 1823, *255–56*; regulations, 1832, *257*; regulations, 1856, *258*; scientific curriculum 1815–1820, 68–74; scientific curriculum ca. 1825, 77–105, 289n9; Second Class instruction ca. 1825, 84–90; sectionalism threat to, 207–8; strategy instruction at, 200–204; Thayer system of cadet selection, 64–68
U.S. Military Philosophical Society, 36–37
U.S. Military Railroad service, 224

Vauban, Sébastien le Prestre de: on fortifications, 4–5, 13, 24, 26, 33–35, 42, 55, 92–94, *94*, 96–99, 227; on mathematics, 4–5; on siege craft, 4, 24, 40, 151, 201
Voltaire, 84

war, art of, 41, 140, 206, 252, 278n12
War of 1812, 2, 11, 39, 45, 48–51, 109, 130, 173, 187, 190, 285nn8–9, 295–96n11, 302n5
Warren, Gouverneur K., 194, 228, 240
Warren, John, 48
Washington, George, 17, 21–24, 26–29, 31, 39, 108, 187, 247, 282n39, 287–88n30
Wayne, Anthony, 22, 31, 108, 282n39
Weigley, Russell, 9, 156
Weir, Robert, 146
Western Department, 133–35
West Point. *See* U.S. Military Academy
Wheeler, J. B., 251–52, *252*
Whiskey Rebellion, 31
Whitney, Eli, 283n53
Williams, Harry T., 166, 168
Williams, Jonathan, 35–37
Winslow, Edward, 18
Wood, Thomas, 163
Wool, John, 72, 162, 166–67

Yale University, 114
Yorktown VA, 24, *236*

Zoeller, Christian, 60

STUDIES IN WAR, SOCIETY, AND THE MILITARY

Military Migration and State Formation: The British Military Community in Seventeenth-Century Sweden
Mary Elizabeth Ailes

Managing Sex in the U.S. Military: Gender, Identity, and Behavior
Edited by Beth Bailey, Alesha E. Doan, Shannon Portillo, and Kara Dixon Vuic

The State at War in South Asia
Pradeep P. Barua

Marianne Is Watching: Intelligence, Counterintelligence, and the Origins of the French Surveillance State
Deborah Bauer

Death at the Edges of Empire: Fallen Soldiers, Cultural Memory, and the Making of an American Nation, 1863-1921
Shannon Bontrager

An American Soldier in World War I
George Browne
Edited by David L. Snead

Beneficial Bombing: The Progressive Foundations of American Air Power, 1917–1945
Mark Clodfelter

Fu-go: The Curious History of the Japanese Balloon Bombs
Ross Coen

Imagining the Unimaginable: World War, Modern Art, and the Politics of Public Culture in Russia, 1914–1917
Aaron J. Cohen

The Rise of the National Guard: The Evolution of the American Militia, 1865–1920
Jerry Cooper

The Thirty Years' War and German Memory in the Nineteenth Century
Kevin Cramer

*Political Indoctrination in the U.S. Army from
World War II to the Vietnam War*
Christopher S. DeRosa

In the Service of the Emperor: Essays on the Imperial Japanese Army
Edward J. Drea

American Journalists in the Great War: Rewriting the Rules of Reporting
Chris Dubbs

America's U-Boats: Terror Trophies of World War I
Chris Dubbs

*The Age of the Ship of the Line: The British and
French Navies, 1650–1815*
Jonathan R. Dull

American Naval History, 1607–1865: Overcoming the Colonial Legacy
Jonathan R. Dull

*Soldiers of the Nation: Military Service and Modern Puerto Rico,
1868–1952*
Harry Franqui-Rivera

*You Can't Fight Tanks with Bayonets: Psychological Warfare against the
Japanese Army in the Southwest Pacific*
Allison B. Gilmore

*A Strange and Formidable Weapon: British Responses to
World War I Poison Gas*
Marion Girard

Civilians in the Path of War
Edited by Mark Grimsley and Clifford J. Rogers

*A Scientific Way of War: Antebellum Military Science,
West Point, and the Origins of American Military Thought*
Ian C. Hope

Picture This: World War I Posters and Visual Culture
Edited and with an introduction by Pearl James

*Indian Soldiers in World War I: Race and Representation
in an Imperial War*

Andrew T. Jarboe

*Death Zones and Darling Spies: Seven Years of
Vietnam War Reporting*
Beverly Deepe Keever

*For Home and Country:
World War I Propaganda on the Home Front*
Celia Malone Kingsbury

I Die with My Country: Perspectives on the Paraguayan War, 1864–1870
Edited by Hendrik Kraay and Thomas L. Whigham

North American Indians in the Great War
Susan Applegate Krouse
Photographs and original documentation by Joseph K. Dixon

Remembering World War I in America
Kimberly J. Lamay Licursi

*Citizens More than Soldiers: The Kentucky Militia and
Society in the Early Republic*
Harry S. Laver

*Soldiers as Citizens: Former German Officers in the
Federal Republic of Germany, 1945–1955*
Jay Lockenour

*Deterrence through Strength: British Naval Power and
Foreign Policy under Pax Britannica*
Rebecca Berens Matzke

Army and Empire: British Soldiers on the American Frontier, 1758–1775
Michael N. McConnell

*Of Duty Well and Faithfully Done: A History of the
Regular Army in the Civil War*
Clayton R. Newell and Charles R. Shrader
With a foreword by Edward M. Coffman

A Religious History of the American GI in World War II
G. Kurt Piehler

The Militarization of Culture in the Dominican Republic, from the Captains General to General Trujillo
Valentina Peguero

Arabs at War: Military Effectiveness, 1948–1991
Kenneth M. Pollack

The Politics of Air Power: From Confrontation to Cooperation in Army Aviation Civil-Military Relations
Rondall R. Rice

Andean Tragedy: Fighting the War of the Pacific, 1879–1884
William F. Sater

The Grand Illusion: The Prussianization of the Chilean Army
William F. Sater and Holger H. Herwig

Sex Crimes under the Wehrmacht
David Raub Snyder

In the School of War
Roger J. Spiller
Foreword by John W. Shy

On the Trail of the Yellow Tiger: War, Trauma, and Social Dislocation in Southwest China during the Ming-Qing Transition
Kenneth M. Swope

Friendly Enemies: Soldier Fraternization throughout the American Civil War
Lauren K. Thompson

The Paraguayan War, Volume 1: Causes and Early Conduct
Thomas L. Whigham

Policing Sex and Marriage in the American Military: The Court-Martial and the Construction of Gender and Sexual Deviance, 1950–2000
Kellie Wilson-Buford

The Challenge of Change: Military Institutions and New Realities, 1918–1941
Edited by Harold R. Winton and David R. Mets

To order or obtain more information on these or other University of Nebraska Press titles, visit nebraskapress.unl.edu.

www.ingramcontent.com/pod-product-compliance
Lightning Source LLC
Chambersburg PA
CBHW031558050925
32178CB00032B/204